"The author introduces a whole new way of thinking about fine arts, design and culture, and business success. This book is a big blow to stagnant categorizations and instead explores how the world is actually operating – by engaging connections rather than mere differentiations." – **Gary Chang**, *Architect, Managing Director, EDGE Design Institute Ltd, Hong Kong*

"Forty years after Andy Warhol taught us that business can be an art, Marco Bevolo shows us how art can make business be better. Intriguing, inspiring and eye-opening: *The Golden Crossroads* is like a piece of marketing art itself. A true manifesto for the post-marketing era, where selling products means creating purpose." – **Wolfgang Schaefer**, *Chief Strategic Officer SelectNY, Berlin/New York*

"What is behind the creative minds of artists? What does a successful cooperation between art and business look like? In times of crisis art producers and art consumers stand on the same side for the first time. Predominant 'MBA culture' has proved we need to change. Noticing various aspects of cultural, political, and social changes, Bevolo calls for new directions in business and cultural liaisons. *The Golden Crossroads* is not a guide, it is not a catalog of good projects – it is a source of inspiration and not only for business leaders, this book is highly valuable for a wide spectrum of readers." – **Anna Krenz**, *artist, architect, critic in Berlin, and chief editor of* Design Vox*, Warsaw*

"Modern, relevant and with an authentic international perspective, this book provides a much needed alternative to the rigid and constrained thinking that hampers innovation. Bevolo demonstrates, with specific and real world examples, how to create the right kinds of bridges between marketing and fine arts, but also between one creative sphere and another. An ambitious and successful dissection of the convergence between art and commerce. Finally, we can see a path towards brands that have true soul." – **Amanda Spring**, *Managing Partner, Janou Pakter Europe, Paris, France*

"For all leaders out there please read this book, open your mind to the next level of thinking and start applying it to your own company and life. Einstein once said 'Problems cannot be solved by the same level of thinking that created them.' This book is about next level thinking." – **Edson Williams**, *founder and director of the artist group Edsonwilliams, London/Amsterdam, and founder and trainer-coach at Lead by Example*

"Don't invite Marco Bevolo for even a coffee if you don't want to have your views shaken. Bevolo takes the stand and speaks his mind. I imagine Bevolo can probably create trouble in the corporate arena by bringing powerful content and creating doubts about the way business is usually carried out. He takes risks and this book is a true materialization of his intellectual adventures. *The Golden Crossroads* is an invitation for us to bet on and invest in the possibilities ahead of us." – **Luciana Stein**, *journalist and trend researcher, São Paulo, Brazil*

The Golden Crossroads

Multidisciplinary findings
for business success
from the worlds of
fine arts, design and culture

Marco Bevolo

palgrave
macmillan

First published 2010 by
PALGRAVE MACMILLAN

Palgrave Macmillan in the UK is an imprint of Macmillan Publishers Limited,
registered in England, company number 785998, of Houndmills, Basingstoke,
Hampshire RG21 6XS.

Palgrave Macmillan in the US is a division of St Martin's Press LLC,
175 Fifth Avenue, New York, NY 10010.

Palgrave Macmillan is the global academic imprint of the above companies
and has companies and representatives throughout the world.

Palgrave® and Macmillan® are registered trademarks in the United States,
the United Kingdom, Europe and other countries

ISBN 978-0-230-22418-6

This book is printed on paper suitable for recycling and made from fully
managed and sustained forest sources. Logging, pulping and manufacturing
processes are expected to conform to the environmental regulations of the
country of origin.

A catalogue record for this book is available from the British Library.

A catalog record for this book is available from the Library of Congress.

10 9 8 7 6 5 4 3 2 1
19 18 17 16 15 14 13 12 11 10

Printed and bound in Great Britain by
CPI Antony Rowe, Chippenham and Eastbourne

CONTENTS

The Golden Crossroads is based on hybrid methodology mixing elements of journalism, qualitative research, and quantitative research. Credits are due to the thought leaders, professionals, and experts who agreed to contribute to the research behind this book. In particular, the author wishes to thank the following qualitative research contributors, who committed their precious time and their valuable energies to share crucial insights to make this book happen:

Anouk Piket, founder, Sichting Caramundo, Rio de Janeiro/Amsterdam

Antonio Tucci Russo, gallerist, Torre Pellice (Turin)

Astrid Wassenberg, curator, TNT, The Netherlands

Atto Belloli, chairman, Istituto Internazionale Studi sul Futurismo, Milan

Concetta Lanciaux, luxury strategy advisor, Paris/Montreux

Davide Quadrio, founder, BizArt, Shanghai/ArtHub, Bangkok

Denise Scott Brown, architect, Philadelphia

Dingeman Kuilman, director, Premsela, Amsterdam

Federica Beltrame, director, Galleria Continua, Beijing

Hendrik Driessen, director, De Pont Museum, Tilburg

Jan Willem Sieburgh, managing director, Rijksmuseum, Amsterdam

Jerry Wind, member of the board, Philadelphia Museum of Art, Philadelphia

Leni Schwendinger, founder, Leni Schwendinger Light Projects Ltd, New York

Lorenz Heibling, gallerist, Shangart, Shanghai

Louise Trodden, director, Art in the Open, London

Maddalena Zolino, co-founder and MD, 515, Turin

Mark Schaffer, gallerist, A La Vieille Russie, New York

Michael Huyser, director, International Fair for Autonomous Design/Art Rotterdam, then: director, Zuiderzee Museum, Enkhuizen (NL)

Peter Carzasty, founder, Geah Ltd, New York

Philip van den Hurk, chairman, Stichting Promotors Van Abbemuseum, Eindhoven, and collector

Richard Meier, architect, New York
Robbert Nesselaar, chief marketing officer, City of Rotterdam
Robert Venturi, architect, Philadelphia
Ruud Visschedijk, director, Nederlands Fotomuseum, Rotterdam
Sandra Khare, director, Chemould Prescott Road, Mumbai
Sarah Schultz, books editor, FRAME, Amsterdam
Sjia Cornelissen, gallerist, Galerie Smarius, Sonnega (NL)
Susan Collins, artist, London
Thomas Widdershoven, founder, Thonik, Amsterdam
Titia Vellenga, head of public relations, TEFAF, Maastricht
Valerie Bobo, founder, Mona Lisa, Paris.

Additionally, equivalent credit should be granted to the generous contributors who enabled the quantitative research chapter of this book by granting their time and professional insights, plus free access to world-class proprietary technologies and to a calibrated US and UK panel of Internet survey respondents:

Alex Gofman, CTO and vice president, Moskowitz Jacobs, Inc., White Plains (NY), who directed and executed the research project

Howard Moskowitz, president, Moskowitz Jacobs, Inc., White Plains (NY)

The Team at Peanut Labs, San Francisco (CA): Peanut Labs is changing the way online market research is accessed and delivered, providing unparalleled access to millions of 13–64-year-olds through social networks. Peanut Labs is the developer of Sample 3.0, its sample methodology that taps into more than 100 social networks including MySpace, Facebook, and Google OpenSocial. More information about Peanut Labs and Sample 3.0 can be found at www.peanutlabs.com. For this book, Peanut Labs provided free access to their sampling capabilities and assets: the author wishes to express his sincere gratitude to them for believing in this project.

The value of accessing people, of gathering unique informal insights and of discussing with thought leaders and experts on face-to-face, relaxed terms cannot be emphasized enough. Among the professionals and experts who implicitly, but nevertheless importantly, contributed to this book in terms of informal dialogs and networking efforts, it is appropriate to mention: Ulrike Erblsoeh, deputy director of the Van Abbemuseum of Eindhoven; Kerstin Niemann, independent curator, Hamburg; Marc Jan van Laake, art consultant in Amsterdam; Donna Wolf, founder of the Deiska' art investment fund; Marline Willekens, collector in Amsterdam; Wilfried Lentz, gallerist in Rotterdam and Alexander Ramselaar, collector in Rotterdam; Piet Hein Eek, designer in Geldrop, the Netherlands; Marko Macura and Ingeborg van Uden, designers based in Eindhoven and Belgrade; Satyendrah Pakhale, "global nomad" and designer between Amsterdam and India; Rogier van der Heide, global director and practice leader at ARUP, Amsterdam; Zaira Mis, gallerist in Brussels; Martin Richman, artist, London; Ron Pompei, founder of Pompei AD and former CNN Principal Voice, New York; Luciana Stein, anthropologist and luxury expert, Sao Paulo, Brazil; Giuliano Molineri, formerly head of international relations at Torino World Design Capital 2008; Corinne Poux, director of innovation at Hermès; Roland Heiler, MD at Porsche Design in Austria; Daniela Krautsack, founder of Cows in Jackets and international conference speaker, Vienna; Mark Tungate, luxury-specialized journalist and author, based in Paris; Steven van der Kruit, creative director, fine fragrances, Firmenich, Geneva; Ichikawa Katsutoshi-san, design director at Sony, Tokyo; Rieko Shofu, executive manager at Hakuhodo, Tokyo; Kou Hattori, founder of Droog Design shop in Tokyo; Dr Richard Lee, luxury industry business leader in China and Hong Kong; Lyndon Neri, founder of Design Republic in Shanghai; Bianca Cheng, director for luxury goods at Jade Investments of Beijing; Rami Farook, founder of Traffic design gallery in Dubai, and many more.

The author would like to thank the team at Palgrave for the efficient and effective support in making this book happen.

A special thank you to Keiko Goto Bevolo and Sara Morosini Bevolo. This book is dedicated to Valentin Iaia Assandri, to Clementine Iaia Assandri, to their mother, my first cousin, Laura Assandri, and to the memory of my father, Carlo Bevolo (1930–2004).

Marco Bevolo

The ways of doing business as we knew them for the last decades have proved wrong. This might seem a bold opening statement for a book by a business publisher. It is however the inevitable conclusion to be drawn from the deep crisis of our financial systems that has been experienced in the last few months. This is not just an economic crisis: it is an ideological crisis, and a spiritual one. Just as communism and its utopian theories proved wrong, so did capitalism in its deregulated form. As much as we might pretend to lead the same old lives as we did before the collapse of Lehman Brothers, we are not doing so. We are already in the first hours of a new dawn, just as Soviet citizens were during the days after the fall of the Berlin Wall, and in the months prior to the end of their state. Without any sort of moralism, capitalism for capitalism's sake just did not work for the rest of us. While corporate tycoons hold on to their privileges and governments desperately try to reboot the system as it was, it is time to think about the future. It will have to be a different future.

The world and the times we live in show us every day that one way of thinking, the larger approach to corporate management that we might simplify as "MBA culture," has clearly reached its glass ceiling in financial and especially ideological terms. However, there exists an extraordinary reservoir of wisdom, vision and potential material for business applications in the fine arts. It is up to the avant garde among business people to anticipate the need to change, and enable the new connections that will transport brands and enterprises into the new markets of tomorrow, in resonance with the new, rapidly emerging societal values. Change is what we need. And change does not happen in a vacuum: it happens in culture, the realm where fine arts and design unfold their power. This is where trends can be anticipated; this is where business leaders and marketers can envisage what will matter to people tomorrow, or a decade ahead.

The Golden Crossroads was originally conceived as a simple analysis of the art market and art marketing. Due to the exceptional world circumstances at the time of its development, it turned out to be an effort to develop a new synthesis, aiming at thought leadership. After at least two decades of one school of thinking in terms of corporate management, ways of working and how business is done, *The Golden Crossroads* has the ambition to nurture a set of potential new directions. It does so by looking at the field of fine arts, design and culture as an alternative source of inspiration for ways to work. It explores diverse visions and approaches in these other worlds. Not each and every direction is meant to be strictly "right." This is ultimately a book that points a finger to the moon to show the possibility of a different landscape.

The key challenge of such a hybrid project in both methodological and content terms was to find a way to transform the wealth of insights, ideas and inspirations that the world of fine arts and design can offer into actionable thinking for business leaders and brand marketers. After all, the arts and business can be seen as worlds apart. This was not an easy task, in terms of editorial approach. To focus on long case histories of success in the marketing of fine arts, or the commercialization of high-end design, would have led to yet another business book, just about a different topic. This is not what we (that is, the author and his contributors) felt was needed at this time. I therefore opted for an alternative architecture to structure our stories and our conclusions.

The DNA of this book lies in the minimalist structure of its chapters: the "multidisciplinary finding." These "findings" are the ultimate synthesis of the information I extracted from various research streams. Their principal sources lie in qualitative research: formal interviews, networking contacts and informal dialogs with "people in the know." Within each "finding," a number of specific observations and questions offer an immediate, yet completely open, translation into potential opportunities and challenges for business and marketing. Providing food for thought in the form of questions to our business leaders and brand marketers is perfectly in line with the inspirational philosophy behind our findings. These are not lessons derived by case histories as such: a lesson implies a top-down, almost arrogant attitude, as in "This is what we know, and now you have to learn it." Our findings come in a different spirit. This book was born as a journey in the world of fine arts, design and culture, so our core message is, "This is what we found out exploring these fields. We believe it is very valuable, so we hope it will inspire your new vision of how to do business."

This book offers 35 findings in the first eight chapters, unified by the common source of qualitative research and by the common purpose of offering inspiration to business leaders and brand marketers. Here we present tricks of the trade as learnt from the world of fine arts. This is where those who are the core audience for the book, corporate professionals and others working in the intellectual sphere where MBA rules apply, will find challenges and opportunities. The last three findings, presented in Chapter 9, are different in nature: they are based on statistics and a quantitative review of public opinion. Their purpose is to address questions about the future of fine arts and design from this different angle, and to speak primarily to professionals in the creative industry and cultural sectors. However this last chapter should also be interesting to corporate managers in other sectors, because it reveals more about the hypothesis formulated in the book, and it does so through the reassuring format of quantitative research. It provides a different accent in the book, but it was conceived to follow on from and cohere with the rest of its content.

This book is designed with a very simple and transparent editorial flow. First, straight after this introduction, the "Golden Crossroads manifesto" conveys the essence of our message. This is structured as a call to action for business leaders and decision makers. Its 11 points respond to the need for a new humanist platform to rethink business and the way we work. Content-wise, it is grounded in, and validated by, the extensive research that underlies the book.

Part I then presents the different "ways of working" and methodologies that underpinned our research. The first chapter discusses the various modes of investigation used, from informal networking to quantitative statistics. Chapter 2 might be seen as the first "real" chapter, with a review of different and diverse definitions of what fine arts are, and with a number of "opening statements" to stretch our minds about the field of investigation. This "content side" of the first part complements the methodologies by entering into the actual topics and themes, with opinions, insights and definitions that were extracted from our dialogs with thought leaders and experts.

The distilled outcome from these qualitative research interviews, performed with more than 30 experts, professionals and thought leaders, forms the backbone of the next three parts.

Part II, "Protagonists," reviews the psychological and marketing communication-related mechanisms behind artists' creativity, art movements and "propaganda," including pragmatic references to boutique consulting agencies and cultural foundations that can connect the universes of culture and business.

Part III, "Relationships and museums," addresses the enabling platforms of the fine arts, at both social and commercial levels, dissecting the rites and rituals of communities, collections and the way museums interact with their visitors and with their urban context.

Part IV, "Futures," engages with an exploration of the future development of the fine arts, starting from a review of the state of the art of design, both in its "super luxury goods" version – what might be called DesignArt for collectors – and in its probable evolution as an engine and resource for urban change. The focus then shifts to the regions where the future of the fine arts is likely to unfold, which are, we believe, the Far East, from Japan to China, and then India and Brazil. Finally this part offers a chapter that is intended to be complementary from both the methodological and content viewpoints. Methodologically it is based on quantitative market research; content-wise, it offers a window on the future of fine arts in the United States and the United Kingdom, according to a qualified sample of citizens selected as online respondents. We believe the methodology chosen for this research also represents a pinnacle in terms of thought leadership in its own field. It was reviewed with enthusiasm by Malcolm Gladwell (author of *The Tipping Point*) in the *New Yorker*; it earned its creators several awards, including the Parlin Prize and ESOMAR Best Paper; it is the topic of *Selling Blue Elephants* by Howard Moskowitz and Alex Gofman, one of the 30 best-selling books in the United States, which has been translated into more than 20 languages. When I looked to select quantitative research partners, I naturally opted for Moskowitz and Gofman since they are true thought leaders in their own field of expertise.

All the chapters include separate box features that offer additional insights on one specific topic, selected to complement and integrate the content flow. These separate boxes address themes as varied as individual artists' biographies, the profile of specific art institutions, the nature of different art movements and the psychology of art collectors. They act as open windows in the book, offering integration to the main body of work and enrichment to the reading experience.

In terms of geographical spread, this is a global, multicultural book. As well as the quantitative focus on the United States and the United Kingdom, our qualitative interviews and the desk research performed as a backbone to the project led us to explore cities as diverse as Tokyo, Shanghai, and Rotterdam, and to speak with experts from places as varied as Brazil, India, London, and New York. Of course, there is a particular focus on a couple of specific locations: first, Italy, from where two key historical art movements provide the main content for Chapters 4 and 5; and second, the Netherlands, which is our key focus in terms of design thinking and the status quo. There is a thread connecting these two relatively distant cultures: for example, in terms of fine arts, it could be said that their welfare policies are exactly the opposite, with the Dutch heavily investing in culture as the "creative industry R&D" of the country and the Italians notoriously leaving culture to its own devices. Looking at the design field, we find Turin (Italy) just having ended its year as World Design Capital 2008, and Eindhoven (the Netherlands) just having introduced its bid to become World Design Capital 2012, after Seoul in 2010. Examples from such differently managed countries can offer interesting extremes to study, to derive universal knowledge from. To focus more on these two countries was an editorial choice in terms of building a strong line of continuity and consistency across the book. It was however also a choice dictated by love: these are the two countries where the author spent the last 20 years, one decade per country.

To recap, *The Golden Crossroads* is not a book about ready-made solutions or "one-minute-management" answers: it is instead a book about enquiries and about questions. Ultimately it is all about inspiration. Its background explorations were performed in the various sectors of artists' creativity and art movements, of collectors and museums, of design and new fine art scenes. Its outcome is a number of findings from the world of fine arts for business people, and the essence of its findings is captured in the 11 points of the manifesto. This manifesto is the point of departure and the zenith of inspirational reference for both *The Golden Crossroads* and your own journey towards what we hope will be a better future.

These are exciting times, because these are the times when people will take their destiny back in their hands: people like you and me, the women and the men who are legitimately entitled to a better future. We are at the center of this manifesto, because the fine arts are among the highest forms of expression of the culture we contribute to co-create every day. In recent years, fine arts, and even design, had become complacent with the Wall Street way of life. This will change: it is time for new critical thinking, it is time for new radical art making and it is time for a new start. Business leaders and brand marketers have the opportunity to benefit from the insights, the challenging visions and the disruptive energy of fine arts, design, and culture. We hope this rebirth might happen along the lines of the following manifesto:

1   People are the goal, not the means, of any business enterprise that is worthwhile.

2   The field of culture is where reality unfolds: understanding culture(s) is necessary to understand the universe(s) where any business enterprise operates.

3   Culture is everywhere, from the cathedral to the favelas: immerse yourself in culture at all levels, by all means.

4   People's creativity is crucial to business success: be it customer co-design or employee training, nurturing creativity is necessary every time, with any stakeholders.

5   Stop producing neutral PowerPoint presentations, start generating ideologies, and a vision of the world as it should be, according to you.

6   Stop producing products, start generating stories, and a narrative line that boldly tells the rest of the world who you are.

7   Stop producing products, start enabling and supporting genuine communities, with the greatest integrity and spirit of service.

8   Design your brand for the long term: there are no tactics, take every decision as if its impact was eternal.

9   Design your brand as an incomplete universe: holistic in ambition, co-creative in people's participation, utopian in its reach.

10   Design and maintain the integrity of your brand according to your own vernacular grammars, and let your corporate identity flexibly follow your evolutions in time.

11   The world of fine arts, design and culture is important because it is different and diverse: always nurture difference and diversity in every way you can.

These 11 points capture the essence of what was learnt from the analysis of how the worlds of fine arts, design, and culture are changing. They also emerged from going back to what – within these domains – did not change over time. In times when all we know will likely evolve into something new, something else, some "eternal truths" of fine art, will not cease to exist. We believe that this might help as a point to jumpstart into the different future ahead of us. These 11 "manifesto statements" will return at the end of our journey in the worlds of fine arts and design, as the basis for a recap of a few key examples, stories and cases, and to provide the key conclusions to the book.

The challenge ahead of us is massive: to reinvent the way we do business, regenerate the way we do marketing, and rethink the essence and the meaning of what we do. The worlds of fine arts, design and culture are the humanist platform providing reference and guidance to constructively move on. These 11 manifesto points are all about all positive belief and optimism in what tomorrow will bring. Humans are what we are: with this manifesto we go back to humanism as the alternative to technocracy and the overpowering arrogance of management as we knew it, in order to move beyond and go forward. These are exciting times because each of us can make a difference to create a better idea of what will come next. This manifesto is a product of its times, with the ambition to last as a reference for positive change.

Methodologies and Perspectives

Introduction

Part I presents two very diverse features, in two chapters:

- our research methodological foundation (or the way we studied our field)

- an overview of major content themes (or a preliminary section of content highlights).

The two research questions addressed within this part are:

- What are the research approaches and the ways of working validating this book?

- How can we describe fine arts, with particular focus on selected connections with business?

Chapter 1 reviews how the content for *The Golden Crossroads* was generated through diverse research phases, both formal (qualitative interviews and quantitative experiments) and informal (networking and immersion in the design and fine arts sector). The primary purpose is to establish a solid basis of validation for the work, while keeping a simple approach to explaining the "ways of working" that underpin our research. Chapter 1 also suggests how informal practices like networking and conferencing can be incorporated into the market research mix. This is a rather journalistic approach to investigating a research theme in the field. In much of my earlier work on corporate research direction I advocated alternative approaches to market research. This is not a book that deals with the theoretical or technical levels of market research, but it might be interesting for research professionals to consider this as one of *their* dedicated chapters in the book.

Chapter 2 introduces some key themes. We look at some definitions of fine arts, and how the field is defined by means of various conventions. These are themes that are developed further in the rest of the book. For example, throughout the book there are references to the biography of the artist and the socializing process, the relationships between art and the luxury industry, and the function of culture and fine arts in the branding of cities. A particular theme of Chapter 2 is connections between the worlds of business leadership and brand marketing, and those of fine arts and design.

In Chapter 2, we meet personalities such as the successful museum directors Thomas Krens and Charles Esche, helping us to understand different ways to put in motion this crucial task of museum management in the art system. Conceptual artists such as Maria Eichorn, Julien Previeux and the Neue Slowenische Kunst collective illustrate how business rituals and structures can be critically analyzed, leading to reflection and reconsideration of our everyday reality. We also discuss how fine arts can offer brand marketers sponsoring opportunities. Finally, we learn from the work of leading architect Winy Maas, founder of MVRDV (Rotterdam) and from the world class 'Luci d'Artista' festival (Turin, Italy) what design and fine arts can do for city beautification. The case study features Fondazione Prada as a focal point of fine arts for luxury marketing and as an urban branding asset for Milan. Chapter 2 introduces some important milestones in terms of perspectives on fine arts, and the book's structure of five "findings" per chapter.

Here are the next ten "stations" in our path through Part I:

- Way of working 1: Immersion and networking.

- Way of working 2: Participation and action.

- Way of working 3: Qualitative research.

- Way of working 4: Quantitative research.

- Way of working 5: Desk research and bibliographic review.

- Finding 1: Expanding our notions of fine arts.

- Finding 2: Expanding on the touchpoints between art and business.

- Finding 3: How contemporary art expands our understanding of the business world.

- Finding 4: Expanding our understanding of how culture works for brand marketing.

- Finding 5: Expanding our understanding of how fine arts support urban branding.

By the end of Part I, readers should have gained a consolidated overview of the journalistic reporting and market research techniques behind the book, and an overview of the topics and lines of thinking that form its content.

CHAPTER 1

The approach taken in this book

This chapter offers a concise introduction to how this book was born. I explain the principles that guided me in my navigation of the worlds of fine arts and design, and introduce the thought leaders, contributors, and partners who made this book possible. The diverse qualitative and quantitative research approaches adopted are presented concisely and effectively. The more immersive, participative ways of working, in terms of networking and so on, are also introduced. The aim is to provide a simple explanation of what was done to discover the world of fine arts, design, and culture. This is not an academic book, as this chapter will make apparent. However it does have a specific ideological orientation, in both research theory and theoretical terms: the principles and the ideas that underpinned the creation of this book are intrinsically part of its message. From this viewpoint, business leaders and market research professionals should find stimuli and inspirations to enable them to reflect on their current practices and on their possible future methodological developments.

The methodological question behind this chapter

What are the research approaches and the ways of working that validate this book?

An introduction to the "ways of working"

As we start our journey into the world of fine arts and design, it is important to appreciate the solidity of the research that identified and validates the book's content. *The Golden Crossroads* was conceived as a visionary book for business people. Its aim is to plant, nurture, and ultimately harvest – for business leaders and brand marketers – a number of new, diverse, and different thinking areas. The challenge was multifaceted, and led to various ways of working:

- *immersive explorations*: attendance at key events and one-to-one networking with important stakeholders

- *action-oriented activities*: direct contribution to key conferences, forums, and symposia

- *qualitative research*: 31 interviews with fine arts, design, and culture sector experts

- *quantitative research*: conjoint analysis with feedback from more than 400 online sample respondents

- *desk research and bibliographic review*.

This chapter dissects each of these research methods and how they were used. It details the "how," the "who," and the "why," and describes the research choices made. At the same time, as *The Golden Crossroads* is a book of creative and innovative thinking, this chapter introduces market research leaders to the challenge to widen, expand, and even redefine their collection of tools and methods.

Way of working 1: Immersion and networking

This "way of working" is based on very simple principle: If you want to understand a field of study, you should be there – in the field – and be part of it as much as possible. This is a technique of investigation that is closer to journalistic reportage than actual market research. It is the approach of the reporter going into the field and observing it from within. In times when market research is challenged to prove its value, the destructured but ultimately sharply focused approach of the great masters of field journalism could be a source of inspiration to both the client side and agencies involved in the corporate market intelligence sector.

To support this way of working, I attended and directly experienced a number of art events, including gallery show openings, leading museum social programs, biennales and triennales, from Venice to Tokyo, and art and design fairs such as TEFAF (Maastricht), FIAC (Paris), Art Cologne, Art Rotterdam and its design sister show, the Dutch Design Week in Eindhoven, and the Salone del Mobile in Milan. I made these visits mostly as a special guest, and sometimes as a member of the press, which enabled me to gather multilayered impressions, from both professional and visitor viewpoints.

In parallel, I developed a number of discussions which did not have the structured framework of a formal qualitative research interview question-naire, but were based on lively and extremely fruitful interactions. These discussions were important in providing personal referrals and flexible advice. The value of accessing people, of gathering unique informal insights, and of discussing with creative thinkers and experts on face-to-face, relaxed terms cannot be emphasized enough.

Networking is one of the most ancient, most generous, most inspiring arts of human relationships. It is an operational mode that business leaders practice in very sophisticated forms, from club membership to informal welfare initiatives in cities and in local regions.

Networking is also important as a research asset: it is a largely unstruc-tured but creatively unique opportunity to define and discover the field of study by being there. This is not the same as classic market researchers: their status of agents that come from the outside world offers them the opportunity to keep a certain degree of distance. This can lead to the dis-covery of new angles and even new foundations in the field of study. It does not, however, enable direct participation. In the last decade, the availability of excellent research suppliers and of pre-digested research reports increasingly translated into the end result of corporate executives and strategic management relying just on data and details from reports. At the very opposite to such sterile ways of working, participation in the form of industry immersion and networking is a powerful alternative in terms of developing an insight into the field that goes well beyond "understanding," and aims at "knowing." It is even better if immersion turns into direct and active contribution to some activities in the field.

Way of working 2: Participation and action

The "Way of working 2" can be considered to be even less orthodox than networking by conventional market researchers and corporate managers. This is where the researcher takes action, and becomes a performing agent in the field of observation. Such an approach to research was devel-oped and defined in the Anglo-Saxon context a few decades ago, with a particular focus on education and its challenges. It is also an approach advocated by the likes of Richard Slaughter, "godfather" of the masters program in future studies at Swinburne (Australia). This is a vision of research adopted and advocated by trend researchers and social scientists with a more progressive, sometimes liberal, orientation. This is the approach of all researchers who do not believe that humans should be observed with the surgical indifference that scientists would dedicate to bacteria or insects.

In the case of *The Golden Crossroads*, I could perhaps claim that I started this particular research phase some 15 years ago, with a leading editorial role at Giancarlo Politi Editore, the publishers of the world-renowned art magazine, *Flash Art International*. This was a unique opportunity to learn not only the glamorous patina of the art world, but also the way this world works behind the scenes. It was also the place to develop a better understanding of the rites and rituals that constitute the informal backbone of the fine arts, where networking circles are vital.

Prior to the conception of this book and over the last ten years, my personal, passionate commitment to the fine arts and cultural sector translated into a not-for-profit contribution as advisory board member to a number of institutions, including the Istituto Internazionale Studi sul Futurismo (ISISUF) in Milan, Platform 21, the incubator of the next Amsterdam design museum, and the Caramundo Foundation (Rio de Janeiro and Amsterdam), and pro bono consulting for art galleries and museums in the areas of brand strategy and communication. This stream of activities was complemented by independent cycles of lectures and conferences on modern art and museum marketing in various settings, such as the Domus Academy (Milan), the C3 Gallery (New York), the University of Leeds (UK), and the trendsetting StreativeSalon events hosted by Shari Swan in Amsterdam. It was of course a unique privilege to engage in such rich extracurricular programs over the years, and one benefit is my insider's view of the field of research for this book.

Later, I undertook a number of activities specifically in order to consolidate, enrich, and validate the ideas and theories presented in the book. Beyond the networking and informal discussions mentioned in "Way of working 1," this recent work included:

- undertaking speaking engagements on topics connected to this book in world-class conferences and symposia – the Art in the Open forum held in late 2008 at the ICA (London), the Cannes Advertising Film Festival, and events in Latin America, Singapore, and the Netherlands

- moderation of a number of panels in the fine arts and design sector – the opening event at Art Rotterdam 2009, the Dutch Design Forum at the Milan Furniture Fair 2009, and key contributions on creativity and the high-end markets at the ESOMAR Asia Pacific conference 2009 (Beijing)

- a cycle of journalistic interviews conducted for magazines such as *Research World* (the official publication of the World Market Research Association) with authors, experts, and thought leaders in the fine arts, cultural, city branding, design, and high-end sectors.

In Way of working 1 (immersion and networking) it is to some extent a benefit not to show or express your own opinions. The ideal networker surfs each different social context with some degree of detachment, and is prudent in espousing opinions and positions. After all, the key role of networking lies in connecting the different and diverse actors of various sceneries and industries: why would a networker run the risk of alienating some people by taking strong points of view? However, a speaker on stage and/or a panel moderator is always actively involved in the discussion. Neutrality is the opposite of choosing where to speak, who to invite to join you on stage, and of course, what to say in public. This "way of working" involves taking a position and a precise standpoint in the field of analysis.

The Golden Crossroads is a book of opinions, and people do not form opinions just by means of detached research. To some extent, they are formed by provoking debate, delivering statements, and actively participating. In writing this book it was important to observe reactions from public audiences, from a conference stage or from the feedback to a magazine article. This is where discussion with professionals operating in the field turned from neutral information gathering into inspiring opinion forming. Of course, just like immersion and networking, this research operating mode was rather ad hoc, unstructured, and open to improvisation. Although I recommend informal ways of working to corporate researchers and brand marketers, there is also a need for more rationally designed research to complement, express, and systematize intuition within a structure. For this book this came in the form of two waves of activities, one qualitative and one quantitative in nature.

Way of working 3: Qualitative research

Complementary to the creative yet unstructured ways of working described above, a more formal, stringent, and repeatable process of research was set up in order to ensure the necessary rigor and scientific soundness of *The Golden Crossroads* as a business book. The qualitative side of this process entailed a round of more than 30 calibrated interviews with thought leaders, trendsetters, and experts in the various fields investigated. I deliberately chose this approach to maintain a line of coherence by involving thought leaders and some less known but insightful experts.

The structure of qualitative research interviews was fixed and repeated across the entire survey. Each question was placed in a "context," presenting the idea behind the specific point, then the specific "question" was addressed in order to expand the viewpoints and opinions of the respondent. The survey was designed to maintain a good degree of flexibility and specificity, while ensuring consistency across the entire exercise. In particular:

- *Variable elements*: In each interview there were three to five central questions. These were created on the basis of the specific CV, profile, and interests of each individual respondent, with the purpose of achieving a profound and structured discussion.

- *Fixed elements*: The first and the last questions of each interview were fixed: formulated and asked to each contributor in exactly the same fashion, leading to consistent feedback on two specific themes:

Opening question:

- *Context*: The departure point of this book is the point of connection between fine arts and business practices. Between these two worlds there have been and are interdependencies and divergences, fatal attractions and ultimate repulsions.

- *Question*: How do you see the general dialectics between the contemporary worlds of fine arts and commercial business evolving at the level of exchange of models and learning opportunities, now and in the next half decade?

Closing question:

- *Context*: One of the points of this book is the departure from the classic notion of contemporary fine arts into the new and unknown. Looking at conceptual openings, this might translate into the birth of a new relationship with the artifact itself, as demonstrated by the rise of DesignArt. On the other hand, looking at geographical openings, we see the whole continent of Asia, from Japan to China, pursuing its own aesthetic and commercial fine arts grammars, which might at some point redefine the rules of the game.

- *Question*: My concluding question to you relates to the possible next phase of evolution of the art system. What do you think will come next, that will move us all beyond where fine arts is right now?

Contributors were offered a flexible approach to their one-to-one responses in the light of their time and travel constraints. This included:

- face-to-face sessions, with the physical presence of both interviewee and interviewer in the same room for some 45 to 180 minutes, with discussion sometimes ranging beyond the interview itself but always ensuring that data collection would be uniform

- teleconference/videoconference sessions, with remote live discussion conducted on the basis of the same principles and rules as the face-to-face sessions

- written response, with contributors receiving their own customized questionnaire in their inbox, and returning it completed.

The flexibility in collecting the responses enabled a wider than planned engagement with respondents. The original expectation was to limit formal interviews to around 10 to 12 people but eventually there were more than twice as many contributors. Respondents to the survey included a great variety of competences and strengths:

- *Consulting firm leaders who managed to connect the worlds of culture and business successfully.* Their professional success enabled them to offer unique insights and experience-based opinions. This subgroup included Concetta Lanciaux, Valerie Bobo, Peter Carzasty and Davide Quadrio.

- *Design professionals with a deep involvement in the fine arts and cultural sectors.* These creative industry professionals managed to use their connections, networks, and intuitions in the field of our study to perform successfully across a variety of industries and categories, from commercial to cultural, from publishing to brand marketing. They included Maddalena Zolino, Thomas Widdershoven, Leni Schwendinger, and Sarah Schultz.

- *Museum management leaders.* The central role played by museums as one of the driving institutions in the world of fine arts and design meant that it was very useful to involve Jerry Wind, Jan Willem Sieburgh, Ruud Visschedijk, Hendrik Driessen, and Philip van den Hurk.

- *Founders and leaders of art foundations.* Alongside museums, art foundations have emerged in the last 20 years as a key driver within the fine arts sector. I chose to focus on the not-for-profit side of the art foundation universe, interviewing Atto Belloli, Anouk Piket, and Dingeman Kuilman.

- *Gallerists, curators, and art fair representatives.* Moving from the more institutional side to the commercial side of the fine arts universe, a mix of professionals was identified on the basis of their insights into where the art market is going: Antonio Tucci Russo, Federica Beltrame, Mark Schaffer, Lorenz Heibling, Sandra Khare, Sjia Cornelissen, Astrid Wassenberg, Michael Huyser, and Titia Vellenga.

- *Architects, urban culture, and city branding professionals.* There were two reasons for involving this subgroup of top talents and experts. First, there was the necessity to represent architecture properly on the panel in terms of its influence on both museum buildings and urban design in general; second, this selection was all but dictated by the rapid evolution of trends following the financial crisis in the fall of 2008. Within the cultural world, a rather sudden (although not totally unexpected) focus change occurred, from autonomous art and design to more socially driven urban design and art interventions (this is discussed in detail later in the book). This was the rationale for inviting the likes of Richard Meier, Robert Venturi, Denise Scott Brown, Robbert Nesselaar, Louise Trodden, and Susan Collins.

The diverse outcome of these interviews was analyzed and reclustered, to form the backbone of the next seven chapters. While the answers to the first and last "unified" questions were compared and aggregated, the flow of opinions extracted from the customized parts of these interviews inspired the soul of this book – up to the point of constituting the starting basis for our quantitative research module, focusing on the future of fine arts as envisioned by the statistically measured opinion of a balanced sample of ordinary citizens from the United Kingdom and the United States.

Way of working 4: Quantitative research

The Golden Crossroads offers new, visionary thinking from the world of fine arts to corporate leaders and brand marketers, thanks to ad hoc, exclusive research. We chose to further verify, validate, and ultimately extend the scope of this journey in order also to offer inspiration to fine arts and cultural sector professionals. It was decided to address the future of fine arts and design, and to engage in this research track by using the more than 30 expert interviews (plus, of course, the implicit knowledge gathered by immersion and networking in the field). The scope of the qualitative side of this book was global, and specifically geared towards thought leaders and experts. For the quantitative section we decided to focus on the United Kingdom and the United States, and to engage with ordinary people. The sampling technique aimed for a mix of education, demographics and other characteristics, in order to ensure the respondents were representative. We asked respondents, "Where are the worlds of fine arts and design going in the next half decade?"

In order to build a complementary viewpoint to the main body of this book, I chose a specific quantitative research approach, Rule Developing Experimentation (RDE) by Moskowitz Jacobs, Inc., of White Plains, New York. RDE is a conjoint statistics method based on a number of simple steps:

- *Understanding the problem and identifying the features for research design*: Here, the research question is clearly defined and a number of variables (or items) are generated and then clustered in conceptual silos capturing their overall essence. This is a critical step, and this is where the qualitative interviews with our experts were best used, by creating a basis of powerful statements and insights representing their collective opinions on possible futures for fine arts and design.

- *Mixing and matching the elements thanks to a specific experimental design*: This step enables an ideologically neutral and statistically sound appraisal of the response of the sample to the different variables by combining them in conceptual prototypes of immediate impact. This step is done automatically, thanks to Moskowitz Jacobs' sophisticated stock of proprietary technologies (ad hoc algorithms and software). Here, the content extracted from expert interviews was combined in a number of sketches or concepts. Each of them was then rated by respondents in terms of how well they felt they answered our question on the future of fine arts and design.

- *Engaging people by exposing the concept prototypes to the selected sample*: This step was performed as a web survey with more than 400 respondents. Here, the sample selection is obviously crucial, as any market researcher knows. Moskowitz Jacobs chose Peanut Labs, the specialized sample provider that uses social networks, thanks to its

Sample 3.0 proprietary methodology. For this project, Peanut Labs selected the specific sample of qualified US and UK contributors.

- *Analyzing, optimizing and identifying patterns and segments*: In this part of the process, automated software and superior statistics led to the definition of what people prefer in terms of both individual variables (or items), and public opinion, as highlighted by aggregating the data into segments. This is where our ad hoc research project identified future directions for fine arts and design according to newly defined segments, and then magnified them from different viewpoints, from gender to nationality to age cohorts. Alex Gofman, the co-author of the bestselling *Selling Blue Elephants* and CTO at Moskowitz Jacobs, generously lent his talent to this key phase.

- *Applying the generated rules to invent the future*: RDE offers the opportunity to understand people in terms of an "algebra of their mind." This means that the experimental process can theoretically go directly from understanding the field of research to testing solutions and visions, with actionable deployment for organizational change and marketing innovation.

This last milestone in the RDE process is also the next step that is awaiting fine arts, design, and cultural sector professionals. It will be up to them to take the conclusions in this book about the future of their universe in the United States and the United Kingdom, and decide how to respond, or even better, how to take the lead in fulfilling people's future expectations. Since *The Golden Crossroads* is a book about vision, it would be excellent to inspire not only corporate managers and brand marketers in the commercial world, but also the professionals who manage the systems of fine arts.

Way of working 5: Desk research and bibliographic review

This last method used for the research on which this book is based is an obvious one: desk research and bibliographic reviews are always performed, especially for this type of research topic. There is an extensive bibliography on the business side of fine arts, theories, case histories of museum management, and so on. However it is worth highlighting that the bibliography for *The Golden Crossroads* does not only include sector-specialized publications and titles. In the spirit of enterprise and vision that is the soul of this book, I consulted a broad range of titles, from the novel *Utz* by Bruce Chatwin to journalistic reportage on the collectors of erotica. The bibliography for this book should be considered not just as a record of what was consulted to inform its content, but as the departure point for new explorations in the various and different fields of knowledge that will, I hope, make this book not only an agenda for action but also appealing and interesting.

Wrapping up: From methods to wisdom

This chapter has provided an overview of the different ways of working that led to the conception, creation, and formal validation of this book's content. According to my professional experience, implied and informal forms of knowledge gathering are key to the research backbone of this work, and to people research in general. I chose to combine a main line of qualitative research with an important ad hoc experimental project in the quantitative side of statistics and samples. The final outcome might look rather complex in methodological terms but it was actually a very simple process, in terms of combining all the five different ways of working as described above. It was simple because it was driven by a clear vision and by the genuine desire to define new relationships in the field for the benefit of business leaders, brand marketers, and cultural sector managers.

CHAPTER 2

Diverse perspectives on the fine arts

The realm of fine arts is a very complex one to examine and define, as it involves the dimensions of aesthetics, economics, social sciences, and much more. This chapter offers a number of easily digestible ways to describe this sector, as an inspirational basis for business leaders and brand marketers. This is therefore a chapter of definition, to enable a first orientation. Here, a first basis is laid to highlight a number of fundamental relationships in the field of analysis. From this viewpoint, key questions are addressed in terms of what fine arts ultimately is as a domain of investigation, some of its possible relationships with business, and how it is interdependent with the trendsetting scenes of leading cities. From this very first "analysis" chapter, connections are made that might appear unusual. However they should also be thought-provoking and inspiring.

The business question behind this chapter

How can we describe fine arts, with particular focus on selected connections with business?

An introduction to the findings in this chapter

Looking at our field of investigation from a purely historical viewpoint, we could observe that, to our primitive ancestors in the caves, their primitive paintings of animals were not what a watercolour or an acrylic painting is to us today. Void of any market relevance, as there was no market as we know it, to those minds, such paintings represented a form of (pseudo-)religious invocation, and a way to exercise some degree of pre-emptive animistic power over future prey. Art then was not art as we consider it, nor was it to some degree in the Middle Ages, when the whole notion of "artist" was rather relative, with religion driving culture as the prevailing ideology. Art changed mostly between the Italian Renaissance and the dawn of the twentieth century. Our modern and contemporary age started with the invention of photography, and the rise of the notion of the avant garde. Within avant-garde movements, art became self-reflective, self-referential, and finally a self-fulfilling field of aesthetic practices and semantic processes.

If we turn to a sociological perspective, fine arts is part of the symbolic uber-structure of any given society at each given moment in time. Fully integrated in the economy of exchange of its own age, art represents the creation, distribution, and validation of those signs and symbols that express and question the balance of power. From this viewpoint, fine arts moved from its ancient purpose and function, that of shamanic and religious rituals, to the self-reflection of its postmodern years – yet always maintaining its function of semiotic representation of societies and culture.

If we move from sociology to an artist-centered perspective, we can identify a different definition of fine arts:

> The fine arts sector is all about how specific people, namely those socially acknowledged and labeled as "artists," are capable of visually formulating their vision of the world and of life, in order to express a view of the reality around us that influences an audience in terms of reflection, and sometimes action.

The next question that spontaneously arises is: "Who is the artist?" The artist can be defined as:

> a visionary with the strength and creative thinking necessary to create their own rules, and with the ability to promote themselves with their peers and with the other constituents of the art world, who contribute to define the field, such as dealers, curators, collectors, and critics.

Of course, this is just a starting point, not a dogmatic conclusion. In fact, opinions greatly diverge:

- To some, artists work according to inner mechanisms of great power, with their own self-defined frameworks and their creativity as the leading factors. To others, artists are geniuses of self-promotion, and individuals with great business skills built into their professional DNA.

- To some, the ultimate artist disregards business; to others, no single marketer can beat the marketing talent of any major artist, from Picasso to Warhol.

- To some, fine arts focuses on the crafts of painting and sculpting; to others it is a domain greatly influenced by the media and technologies of its times. Think of the IT revolution in the last decade, and the influence it played on the creation, production, and distribution of fine art. Or think of the application of sheer marketing techniques to art investment as perpetrated by the likes of Charles Saatchi, the former advertising genius turned mastermind of the art world.

- To some, the commercial field of galleries and auctions is *the* actual field of fine art; to others, the amateur painter in the Australian countryside is much purer and more relevant in terms of existential meaning than the taste-creating artist on show in London or New York's prime spots.

Opinions and arguments could continue because this is all open to debate and to speculation, much as any scholar can frame it into the rigor of a PhD study. Of course, all these definitions are from our present-day culture; in different times, different forms of societal institutions and relationships will affect and sometimes dramatically change the notion of what fine arts is. Ultimately, who is then tasked with deciding what art is, for the audiences of today and tomorrow? Currently, this could be seen as the role of institutions such as museums, and creative leaders such as museum curators and directors. It is up to them to maintain an overview of the entire field of fine arts, and select what truly matters to represent our culture, to both current and future generations. The latter will look at museum collections and milestone publications to understand both the history of fine arts in our societies and our culture through its artistic production.

These introductory notes outline the breadth and the complexity of our field of study. Within the context of this chapter, we address this complexity with five "findings" in response to the following questions:

- How does an artist become an artist, and how does the evolution of media impact the definition of what art is? This first "finding" offers two angles to gather a deeper understanding of the different functioning levels of the fine arts system.

- How do fine arts relate to business in terms of leadership? In this "finding," art meets business, with some insights into the gap between MBAs and visionaries.

- What do artists have to say about the world of business? In this "finding," examples of how fine arts reflect on business processes are discussed.

- How can we understand relationships between fine arts and people? In this "finding," dynamics of interest for brand sponsoring purposes are discussed, including the distinction between the audiences for high art and popular culture.

- What is the relationship between fine arts and the branding of cities? In this "finding," a first overview of relationships between city marketing, design, and fine arts is presented, for the main benefit of those involved in public art.

By the end of this chapter, we will have discovered a few basics about the fine arts sector and its inner workings, while sketching connections with the worlds of marketing, branding, and business. This section does not focus on questions to stimulate marketers, brand managers, or business leaders to think and rethink their practices. That is the purpose of the book as a whole. Instead, as already introduced, the purpose of this chapter is to lay out the framework and background to what follows.

Finding 1: Expanding our notions on fine arts

Understanding the fields of analysis

The introduction to this chapter sketched the territory of fine arts from different viewpoints. In particular, it mentioned how the role of the artist is central in defining what fine arts are all about, and introduced the notion of fine arts as depending on the dominating media landscape. In this first "finding," we further expand on these two important aspects, to understand them more in depth. First, we discover how an artist is "made" by following the standard hypothetical journey of a young talent from art school to fame. This will give us a clear overview of how the protagonists of fine arts navigate their own world. Second, we explore the hierarchies of arts according to media, and the power that the evolution of media technologies has on the formats and the canons of what fine arts is at every given moment in time. Business leaders and brand marketers are invited to read the next paragraphs bearing in mind parallels with their own situation.

How does the fine arts system turn any of us into an "artist"?

If you were an artist at the beginning of your career, it could be greatly beneficial to start from an academic curriculum in the fine arts or design schools. This would give you the opportunity to both learn and network, being acknowledged for your talent by your peers and your professors. Once you were a graduate, you would show your work in galleries and other institutions. First, you would need to choose between commercial art, with its circuit of merchants, agents, and fairs, and public art, with its city commissioners and other liaisons and relationships to be built, such as with governmental bodies and public agencies. These two worlds appear still to be rather apart and not in communication with each other, and not many artists manage to be equally successful on both sides of this "gallery" versus "public" art divide. If you chose the commercial sector, you would need to gain representation by a good primary gallery, one that advocates research and supports new artists. Such a gallery will have a circle of collectors, customers who will have an ideological, aesthetic, or generational affinity with your work. This will mean shows and sales. In parallel, your CV with museums, foundations, and other fine arts institutions will grow as well, shaping your career path.

A number of milestones will represent the actual path to fame: the first solo show in your gallery, the first show in a leading provincial museum, the first acquisition of your work by a leading international museum (the Museum of Modern Art in New York (MoMA) being among the top options). At the same time, the more commercial circuit of fairs and auctions will establish and benchmark your price and your market value. Also, at fairs, collectors and curators will network and gossip, and will talk about your work and you, further propelling your brand in the art world. This is the world described by Don Thompson in his *The $12 Million Stuffed Shark* (2008): a brilliant read which enables you to understand a complex yet coherent system.

Observing and understanding in depth how this works is both a great privilege, and an outstanding learning opportunity, especially for marketers working in premium value categories. The preliminary conclusion for brand marketers in general here is simple: We might think of an artist as a brand that has the possibility of lasting for centuries.

How does the evolution of media play a role in determining "what" is art?

A second way of looking at fine arts is to analyze it from the viewpoint of its media evolution. Historically, many consider the fine arts as just paintings and sculpture, and these two facets still remain crucial in the contemporary system of fine arts. However, there is clearly more in terms of the relationships between new media and fine arts categories. The rise of new media historically impacted upon and redefined the entire field in continuing cycles. The birth of photography in the nineteenth century could have been classified initially as the pure application of technology for applied arts (for example portraits). In reality, photography soon made an impact in the field of fine arts by inspiring painters to move beyond visual realism. This meant the analysis and deconstruction of the mechanics of their technique and their understanding of light, leading to impressionism, divisionism, and other late nineteenth-century art movements. Also, the evolution of photography across the twentieth century made it first into a new discipline with great relevance in applied arts, and then into a privileged medium for the 2000s in terms of actual fine arts. The next steps of its evolution will be determined by institutions such as the Nederlands Fotomuseum of Rotterdam, and are further analyzed in later chapters.

What matters here is how a single medium, photography, moved in a few decades from being just a technique to becoming a true fine arts discipline, with its own right to exist and to be taken seriously by museum curators, collectors, and dealers. In the future, of course, the same opportunity will apply to digital media and any other technological basis for human expression at a visual level.

Here, a potential opportunity of reflection for brand marketers is simple: rethink your media approach, by accepting that new, innovative, and even disruptive media and channels to reach people will change the rules of the game. There has been a lot of talk in the last 20 years about people-centric and nonconventional planning. Anticipating the evolution of media means being where the important and key people are before anybody else, just as photography collectors, dealers, and museums were for a few decades, before today's explosion of interest in this medium.

In conclusion, let me reiterate that fine arts is a field that can be described from different viewpoints: sociology, media studies, history, or the simple governing mechanisms that make it work all year round as a system for enabling artists' careers. From a more commercial viewpoint, of course, fine arts is a business sector with rules, rituals, and systematic ways of working.

Finding 2: Expanding on the touchpoints between art and business

Examples of how distant worlds converge

The relationship between fine arts and commerce is complex. We might fairly say, as we discovered in our first "finding," that marketing is present in the mind of artists as the natural way to relate to their audiences. Of course, not everybody will agree. To some experts, the worlds of fine arts and business are, and should stay, separate, at least in everything involving the act of creation. To those purists, art is a spiritual affair or a philosophical quest. Think of great modern masters like Jackson Pollock or Mark Rothko, and their existential path through painting. On the other hand, looking at the signature names of contemporary art, from Damien Hirst to Takashi Murakami and Jeff Koons, behind their success we find a company structure with a number of contributors and specialists that can number from ten to 100, and with production processes that closely resemble those of a sophisticated workshop of luxury watchmakers in the Swiss Alps or a high-end leatherware maker in Italy.

For the sake of simplifying our argument, we might take on board an "artist equals marketer" viewpoint. After all, we are looking at a huge market in dollar terms, and at a complex system of dependencies regulating the assessment of value for specific items, mostly treated as super luxury goods in microeconomic terms. From this viewpoint, art is just business, perhaps a business sector with different rules than fast-moving consumer goods or retail, but nonetheless with strict rules. Within this vision of the art sector, all actors are purely economic players, regardless of the specific aesthetic or ideological content of their trade or creation. Here, marketing thinking is important to establish the experience of the art audience, in the same line of work as marketers experience in their business practice. Perhaps it is only the complexity of stakeholders and the subtlety of their reciprocal relationships that makes a difference, with artists, curators, sponsors, visitors, public officers, and so many other diverse players at hand. In this respect, the sociocultural status of fine arts collectors and their world points towards a direct connection to a business domain with great economic importance (at least until the financial crisis), the high-end and luxury market.

Let's follow this path and expand our exploration further to look at the luxury industry itself, with its great names of fashion, stylists, and taste makers. This is a business domain where the connection between artists and brands is relevant. According to some experts, a proper understanding of the entire business of high-end and luxury goods strictly requires the study of the fine arts and its history. The reason for this lies in the cultural notion of "style." The concept of "style" is fundamental to the entire evolution of fine arts, from the caves of primitive peoples to the Guggenheim Bilbao or MoMA. The great masters of classic painting and sculpture created their own styles. Today, we would call those signatures brands. What more powerful brands are there than Leonardo or Caravaggio or Rembrandt, which have lasted for centuries? For some among them, this resulted in incomparable commercial fortunes over time, with a success that most brand managers can only dare to dream of. In an ideal world, it is natural to conclude that luxury executives should be properly trained in the history of fine arts – especially in understanding what fine arts can mean to their brands.

I certainly advocate the great potential for the luxury industry to learn from the fine arts. However, we might also engage in a bidirectional discussion between the cultural sector and business management about the fundamentals of leadership. We can find learning opportunities in such an exercise as well. Two examples should show what I mean.

If we were to consider the names of leading museum directors in the world, Thomas Krens, the American MBA who for the last 20 years has led the Solomon Guggenheim Foundation and its museum brand, would probably be in the top five. Krens has a reputation for building a business strategy that led to unprecedented financial investment in the art and museum sectors. He is the business mind behind the Guggenheim Bilbao, and the expansion of the Guggenheim brand into the United Arab Emirates. At the opposite extreme to Krens is Charles Esche, the director of a Dutch avant-garde institution, the Van Abbemuseum. This is a relatively small, 70,000 visitors a year, public museum. With the vital support of his deputy director and business partner Ulrike Erbsloeh in continuing the history of the Van Abbe brand, Esche pursued an aggressive, antagonist vision of what art is, and what function it should play in society. Controversial, bold in his statements and viewpoints, he brought the Van Abbemuseum back to its 1960s political edge, anticipating subjects such as the resistance to global capitalist "unique thinking," addressing issues of citizenship, society, and culture well beyond aesthetics and experience, and staying relevant in an international scene of contemporary art increasingly unified around commerce. Yes, it is easy to suppose that the axis created by these two personalities reflects the polarization of the worlds of MBAs and the world of visionaries.

The easy supposition is that in the near future, the museum world will demonstrate a move from the former to the latter, with genuine visionaries taking the lead. This will of course happen only if such visionaries can deliver their controversial and inspirational ideas with limited budgets.

In conclusion, fine arts is a business sector, but a very unusual one. This makes fine arts a business domain that requires more exceptions than rules, perhaps, but where rules nonetheless do apply. On the other hand, if we move from "process" to "content" and from fine arts in general to contemporary art, we can identify a number of inspirations for business leaders, this time straight from the actual content of artists' work.

Finding 3: How contemporary art expands our understanding of the business world

Exploring alternative viewpoints on business reality

How do art and business enter into an operational discussion in the world of brand marketing? As a clear (if suboptimal) example, let's take the recent Chanel Mobile Art traveling container, designed by Zaha Hadid. This is in many ways equivalent to the commissions powerful people gave to fine artists in the Renaissance. A selection of very good contemporary artists were invited by an outstanding French curator to produce works based on Chanel "designer handbags," which were displayed in Hadid's mobile exhibition space. From the design viewpoint, visiting the container and its setting (I did so when it was in Tokyo) was an excellent experience. From the viewpoint of fine arts, this project is probably as close as you might get to intellectual prostitution. It could be argued that it was just a noncritical injection of fine artists and their works into the Chanel brand experience. It was a remarkable journey for Chanel brand lovers, a rather commercialized way to integrate signs and artifacts produced by artists in a sterilized setting, void of any analysis or critique beyond the almost vulgar self-celebration of the "super-IT bag," which it vainly tried to elevate to art matter. This is pure image making transferring from the world of fine arts into the world of industrial PR.

We should not consider the Chanel Mobile Art project as an optimal example of cross-pollination between business and fine arts. Learning from this particular case, important questions are raised for business leaders and brand marketers:

- How can we benefit from the critical thinking that only true fine arts offer?

- How can we use the creativity of artists and movements beyond simple PR?

- How can we connect the disruptive power of fine arts to everyday business practices?

"Interpretation" is the keyword in the bridging of contemporary art and fine arts in general in any different situation. Crucial to contemporary businesses are characteristics of fine arts such as the relevance of creativity, the absence of a "one way of thinking, one solution" to all problems, and the value of diversity. After decades of MBA-formatted thinking in terms of approach to strategy, the fine arts principle, at best described as: "there are no sacred cows," is a key asset in facing the crisis by seeking profound change. While it has been, and will increasingly be, imperative for art ventures to regard themselves as business enterprises, pursuing lateral thinking as an opportunity to change is crucial for standard business practitioners.

The richness of critical thinking in contemporary art becomes evident when artists express themselves – without sponsoring obligations or any other commercial constraints – on topics and themes that happen to analyze business practices. We tentatively list two approaches here, exemplified through the work of three artists:

- The analysis of business processes and practices as a theme of insightful conceptual work, dissecting the very structures of what business is about.

- The critical analysis of particular relationships of power in the business sector, as seen by an artist who takes a political position of a provocative nature.

Addressing our first point, there are lines of research in contemporary art that investigate the nature of business as we know it. German conceptual artist Maria Eichorn looked into the nature of capitalism in its simplest mode of existence – the shareholder company. Her work includes *Money* at the Bern Kunsthalle (2002). In this instance she utilized her exhibition budget to repair this historical institution's building and to replicate the way it operated at the beginning of the twentieth century, in issuing unlimited shares. This project brought the artist to operate at the very heart of company governance. Also in 2002, as part of her contribution to

Documenta in Kassel, Eichorn created as an art project, Aktiengesellschaft, a not-for-profit company. Again, the constitution and formal administration of what is a real shareholder company becomes an opportunity for fine arts to comment on the basic functioning of business, and to operate according to its very formal rules for the purpose of presenting them in a thought-provoking form.

There are also more direct ways to involve business management in fine arts projects. By coincidence, also in 2002, graduating artist Henrik Schrat organized an interesting "art meets business meets art" project: a "Manager in Residence" program at the Slade School of Fine Arts in London. Schrat invited a business manager to join the school, with the same objectives and operational modes as an artist in residence program would offer. Over a period of six months, the manager was given the opportunity to perform research and analysis, and to generate a vision of the world that became the basis for the creation of an original artwork in the form of a comic strip book. What is interesting here is the direct inclusion within the art context of the wisdom of business on one hand, and the challenge to business thinking to operate according to the standards and procedures of the art sector. Here the reflection that artists might offer addresses the very essence of business in its everyday practice.

Moving our focus to the second point, the critical analysis of relationships in the contemporary world of business, we find French artist Julien Previeux. In 2007 he created an art project, to be published as a book: *Lettres de non-motivation* (Letters of non-motivation). The artist reacted to a number of ads for different jobs in the French market by all kinds of potential employers. He did so by sometimes making fun in a sarcastic fashion of the copywriting formats and the standard formulas that rule the labor market in a sterile fashion. Starting from the language, styling and implied rules of the game embedded in such ads, the artist recreated a number of letters aimed at explaining why he "refused the job offer" as described in the ad. Sometimes with analytical rigor, sometimes with anarchist arguments, always with a subtle sense of humor, Previeux engaged in a "dialog of non love" with HRM departments in his country, leading to sometimes comical, sometimes politically correct, mostly careless feedback from his prospective non-employers. Within this work, Previeux raises the criticism of the relationships and interdependences between demand and supply in the labor market to a whole new level of analysis. His basic materials are not colors or sculpting stone, he molds and models our everyday into his own vision by utilizing the style and the dialectics of vacancy announcements as they are published in French dailies and magazines. The final outcome is a rather provocative example of critical thinking.

If the works I've mentioned by Eichorn and Schrat offered the opportunity to think about general business structures, Previeux challenges business people to get new political insights. He does so by intervening at the very heart of the linguistic methods that companies use to represent their role and to present their demand in the labor market. Why are these artworks important to business leaders and corporate managers? Because they anticipated methods, connections, and turning points that led us to the present crisis half a decade before its arrival. The study of fine arts can therefore offer a rich cultural insight, to those willing to engage deeply with it. This means getting ahead in terms of future analysis and trend studies, and even more crucially, getting the opportunity to reflect and to reconsider your everyday practices.

Finding 4: Expanding our understanding of how culture works for brand marketing

Reaching people's minds, touching people's hearts

It is usual that in times of economic downturn corporations cut their marketing investments. This is a negative approach, since these are exactly the kind of investments that will connect product to people: and if you cannot reach people, why on earth would you ever produce any goods at all? Of course, through the business downturn, most corporations will likely end or substantially reduce their cultural sponsorship too. I believe they should not do so. Reaching and endorsing communities by means of culture is such a strong asset in terms of reputation building, especially in such times. Here, sponsorship would go beyond pure access to future customers: it might instead mean establishing a dialog with markets in the making. This, of course, demands a superior approach in strategic terms, and nuanced programs for reaching people's hearts, not just through classic advertising.

For sake of clarity and effectiveness, here we will oversimplify history and theory. To the Catholic Church, at the time of the Old Masters, art was to some extent a tool to bring the Gospel to the masses. Of course, even then, to some, it was also an opportunity to indulge in beauty and sometimes sensuality. However a clear communicative structure was installed to connect content and people through form and aesthetics. Canons were set on how to represent saints and "heroes of the Faith." Let us assume that this framework was not that different from contemporary advertising and its highly codified campaigns – although there were a different message, different audience, different persuasive methods. The point is, in the past art entailed a deeper communicative purpose. Of course, this is a rather simplified picture. However there is no doubt that with the modern avant garde, all of this changed. Art became the subtle game of pushing the borders of fine arts and high culture, and this left people sometimes divorced from and sometimes puzzled by the actual meaning of what they saw.

The distance between the more intellectual and sophisticated level of high-art production, the one on which Eichorn and Previeux operate, and the actual level of commonly shared popular culture, is effectively demonstrated by the coexistence of both in a collective such as Neue Slowenische Kunst. This is a fine arts and performing arts collective that has operated since the 1980s. It incorporates a visual arm, IRWIN, focused on galleries and museums, and an industrial techno rock band, Laibach, which generates a great amount of political controversy. IRWIN operates through the fine arts system, Laibach manages its distribution across the channels of pop culture. These two entities are connected by the very same vision and insight. However they operationally divide their tasks by reaching different audiences in different terms, almost unrelated to each other. The gap between high art and popular culture is still large. But is it really? There is an important classification in brand marketing and sponsoring terms:

- On the one hand, we have high art, or what goes into museums and is awarded public recognition as the stuff that geniuses create.

- On the other hand, we have popular culture, or the world of entertainment with lower dignity of expression that provides our everyday fun, from radio to reality TV shows and TV drama.

This divide has historically been defined as a function of both the sociological relationships of power and the actual perception of media in societies. In ancient times, for example, poetry was considered a much higher field of artistic expression than sculpture, with the latter being almost confined to applied arts and craftsmanship. A number of fine arts and design movements in the modern and contemporary age have addressed such a divide. They did so by promoting a vision of fine arts with a highly inclusive orientation. Their main purpose was to connect and to deploy fine arts into our common everyday.

A clear understanding of the relationships and hierarchies within culture is of course very important to brand marketers. Here, the framework of distribution and reproduction of a given cultural manifestation becomes a fundamental reference for potential sponsoring purposes. For example, visual arts and classic music represent interesting options for those marketers seeking the ideal combination of cultural prestige and dollar value in PR return. This is where higher income strata and more affluent segments focus their cultural consumption. What more can be done with sponsorship, if we remain in the realm of higher culture?

Event sponsorship is as old as marketing: high-end brands have innovated in this field by taking the lead in recent decades. Luxury leaders have already moved beyond sponsoring third-party events. Here, the art foundation format emerged as the most sophisticated option to combine true cultural prestige with actual PR firepower. Fondation Cartier in Paris, the Shiseido Art Gallery in Tokyo, and Fondazione Prada in Milan remain among the most relevant fine arts hubs in the world, thanks to their programming. However they are also exceptional vehicles of reputation management for the brands that sponsored them. Operating at the very heart of world-class cities in the world of luxury business, and of culture, these foundations represent a unique opportunity. Putting culture in the city and at the heart of their branding proved incomparably valuable for these high-end players. It was also a great way to connect their marketing strategy to the culture and the profiling of the cities where they operate, almost in terms of giving back. In the case study in this chapter we address the history of Fondazione Prada.

Finding 5: Expanding our understanding of how fine arts support urban branding

Creating the ultimate match to profile cities

Urban centers such as Florence, Rome, and Paris were the birthplace of classic art. Urban centers such as London or New York are the drivers of today's contemporary art markets. Cities are historically *the* place where fine arts happen. In the United States, the birth of civic communities was often marked by a building connected to their cultural status – a library or a museum. Nowadays, the presence of culture in cities is a major asset for tourism marketers. Urban institutions of culture have tightened their connections and links to public bodies and corporate investors, and they have become (in their best-case scenarios) part of integrated systems aimed at elevating the urban experience and the city branding profile. We could say that today a city without a serious focus on fine arts and design is not a city to be taken seriously in global terms. In urban marketing, the competition in terms of art biennials, museums, fairs, and all things artistic has become tougher and tougher. Within this context, a city should think in brand strategy terms, by orchestrating its cultural system towards any specific positioning as appropriate: for example, "pioneer of new art forms" as opposed to "home to ancient classic art." This is where former industrial capitals such as Turin or Eindhoven aim to become creative industry hubs and even world design capitals. Just like any other enterprise, a city cannot excel in everything: as with all marketing practices, making choices and targeting investments is vital. Can we identify indications of emerging trends in this field of art, design, and city marketing?

Following the ideas of thought leaders such as Andrea Branzi and looking at our cities from the perspective of urban design, it seems natural to imagine that we do not need, at least in the advanced economies of this world, one more conventional icon, be it an artwork, a bridge, or a museum. The economic downturn brought a whole new focus on functional features within our urban planning priorities. Aesthetics will have to fit and follow function, creatively. Within such trends, the beauty of a LED-decorated infrastructure might make a humble spot of urban design almost into a potential city branding icon. This is the case in Eindhoven, where the municipality made a conscious effort to focus its urban lighting strategy into organic and organized programs. Around the local regional airport an installation such as the Flight Forum by Eindhoven-born architect Winy Maas of the Rotterdam-based firm MVRDV, shows a glimpse of future urban design. Conceived by Maas, the business building entails a simple planning grid. What makes it special, however, happens at night when the blue illumination creates a highly suggestive lunar landscape. The Flight Forum benefits from a lighting solution designed for functional purposes because the blue light aims to reduce drug activity in the deserted business building at night by making it impossible for drug addicts to inject their veins. This is definitely more of a design solution than a fine arts decoration. Nevertheless, the careful planning and the particular aesthetics of the final outcome make this site a perfect example of how architecture and design make an aesthetic difference in urban areas that until a decade or so might have been regarded as purely functional. This new approach to urban beautification by means of aesthetically elevated infrastructure will be one of the keys to moving towards the new creativity that will drag us all out of the economic downturn.

Moving from urban design to fine arts in the city, an example of how culture can help to develop cities is given by the use of urban lighting-based fine arts programs in the public realm. As part of the necessary infrastructure, lighting is inherently democratic and highly pervasive. In the last decade the creation of fine arts events involving urban lighting interventions has started to pick up. One turning point was when the former industrial capital of Italy, Turin, created "Luci d'Artista" (Artists' lights). Planned every year around Christmas, this festival shows the works of world-class contemporary artists from the gallery/museum circuits, such as Daniel Buren, Rebecca Horn, Jenny Holzer, and Jan Vercruyssen. The artworks are installed in the form of lighting within the most inspiring areas in town, from central squares of great baroque elegance to the densely inhabited suburbs and immigrant areas. The result is unique in terms of both quality and integration. Quality is acknowledged by the enthusiastic response by experts worldwide, from London to New York and from Philadelphia to Lyon. Integration within urban design is ensured as well, with passers-by experiencing beauty in its most sophisticated form.

At the dawn of a downturn age, "Luci d'Artista" shows the way forward in terms of public art. It is part of the "ideal portfolio" of inspiration that sets the agenda of public bodies, such as "Art in the Open" in London, part of OpenHouse and connected to Design for London, London 2012, and a number of additional public bodies with an interest in putting aesthetic and cultural quality back into the city. Here, public forums and civic lobbying aim to increase the presence of art in the very heart of London, with an eye on the 2012 London Olympic Games. "Art in the Open" responds to a larger trend, combining integration of fine arts quality in the urban texture and a new way of thinking about the city. It is one vision where fine arts – in this case, public art – is and will be increasingly central at strategic level. Brand leaders and strategic marketers should definitely take note of this strong trend.

Wrapping up: from fine arts to finer business

In this chapter, we first considered a better understanding of what the fine arts sector is about, and how its connections with the world of business can be identified. We looked at how the dialog between business and fine arts is definitely a two-way street at an analytical level. On the operational side, we concluded that the need for sponsorship for the cultural sector will match the equally strong necessity for companies and corporations to enable the growth of critical thinking and alternative visions of the world.

This is because the nature of the current crisis is once again ideological, and not just financial.

The relationships between fine arts and business are multiple and not simple. We observed how the world of fine arts is a very difficult one to define as a field of investigation. We also discussed how the system of fine arts is an economic entity with a great degree of complexity but with a number of standard rules, making it sometimes close, in its nature and its audiences, to the worlds of luxury and high-end goods. We should also highlight one of the vital differentiating elements between the world of fine arts and business, namely the very nature of artists – personalities with some degree of magic aura, whose modes of production and relationship do not reflect the standard rules of marketing. Given their centrality in the fine arts, a more comprehensive understanding of the minds of artists is necessary as the starting point in our quest, and this is the focus of the next chapter.

Artists' works with business as a subject can disclose a number of hidden truths in business itself. The nature of fine arts – analytical, critical, sometimes radical – can help business people to move beyond the confines of their somewhat sterile technocratic education, and their standardized mindsets. In a (business) world that made innovation one of its (lip service) priorities and which recently showed how weak its foundation is, the different viewpoints of artists should be more than welcome. Of course, on the other hand, there is a parallel necessity for museums, commercial dealers, and artists themselves increasingly to structure their operations and strategize their enterprise assets.

When it comes to forming trends, it must be said that the prevalence of urban culture will stay, whatever the crisis. This will be the case especially in those advanced economies that started their reconversion from industrial to cultural drivers during the great prosperity of the last decade. The way art interacts with cities will be vital at more levels: inspiration, education, and the pragmatic beautification of the urban territory in times of low budgets and cost cutting. The most forward thinking, the most advanced ideas, the forming of taste and trends will remain at the heart of cities such as Tokyo, Milan, and Paris, and of course, New York and London. On the other hand, smaller cities, provincial capitals, and centers of excellence such as Turin (World Design Capital 2008) and Eindhoven (a candidate for World Design Capital 2012) do offer a few "nuggets" of advanced thinking.

Chapters 3 through 9 discuss these themes and lines of thinking in further depth. It must be realized that fine arts is not just a field of super luxury goods, at least not after the 2008 demise of Lehman Brothers. Likewise, it must be said that fine arts is not an ivory tower of entertainment, or a space of pure spiritual and philosophical debate. These first five "findings" provided some elementary points to engage discussion. The rest of the book will stretch and challenge the current paradigms and self-fulfilling prophecies of business thinking.

Fondazione Prada: Birth of a fine arts brand

In order to understand Prada, the Italian luxury brand, it is perhaps best to visit its New York store. From stores designed by Dutch architect Rem Koolhaas, to the US headquarters space cleverly converted (by Herzog & De Meuron architects) from an anonymous industrial loft on the Hudson, the whole ethos of the brand lives in the very flesh and blood of those walls and glass. In Milan, Prada has several flagships, starting from the 1910s first shop in the hyper-central, hyper-chic Galleria Vittorio Emanuele where Miuccia Prada's family started selling their high-end travelware to the wealthy of their time. However, the brand focuses on its future rather than in its past. This seems natural, for a trend-setting, taste-making, experimental brand like few others in the arena of fashion and luxury worldwide.

This could be the reason why Miuccia Prada, with strong academic credentials in political sciences, together with her temperamental husband and business partner, Patrizio Bertelli, initiated the not-for-profit Fondazione Prada. Miuccia Prada is noted for her political thinking and her analytical skills. During the 1970s, she disregarded her family business to concentrate on academic studies, radical left-wing politics, and feminism. This cultural background gave her a different taken on her business later on, and an ideal basis for her interest in contemporary art. It must be added that Milan is not a generous city when it comes to contemporary art. Although it was the center of important early 1900s movements such as Futurism, and is the natural commercial and business capital of Italy, the museum system in the city does not match the actual levels of fashion and design excellence. The two major Italian contemporary art institutions are the Castello di Rivoli (Turin) and the Museo Pecci (in Prato, near Florence).

At the time of opening the Fondazione Prada, in the mid-1990s, there was almost no truly international exhibition space in Milan. There was the PAC (Padiglione d'Arte Contemporanea) devoted to local or national-level shows, the MiART fair, also local or national at the best, and of course there was a good system of leading commercial galleries, from Massimo De Carlo to Emi Fontana. Fondazione Prada changed the scene substantially. Officially started in 1995 with a show by prominent sculptor Anish Kapoor, and followed by important retrospectives (such as Dan Flavin) and its own productions of emerging talents (such as Francesco Vezzoli), Fondazione Prada assumed a unique role in Milan. Choosing historical contemporary art guru Germano Celant as the director in charge of the program, Prada made a definite step towards world-class excellence. Miuccia associated her brand with the fine arts signatures of the likes of Michael Heizer, Laurie Anderson, Sam Taylor-Wood, Enrico Castellani, and Mariko Mori. The production of catalogs and publications, and Miuccia's presence in the media with her strong opinions, ensured the ultimate alchemy: a space of cultural innovation at the highest quality of fine arts production, feeding an excellent PR machine to generate relevant value for the brand, with worldwide resonance. Art met business, with integrity and respect, and they both live happily.

Protagonists

Introduction

Part II builds on a number of themes introduced in Chapter 2: the artist as engine of fine arts, the relationships of fine arts and applied arts, design and advertising, and some dynamics of branding connected to contemporary art. It also introduces a number of new themes: most notably, the way an important modern art movement was launched in terms of propaganda techniques, and what this means for us, marketing practitioners. Two questions are addressed in the ten findings of this part:

- What is creativity and how can it work in the business context?

- What can the world of fine arts teach to today's and tomorrow's brand marketers?

Part II has two distinct directions. The first, pursued in Chapter 3 ("Psychology of the arts: how it works for work") is about the "mind of the artist," their inner workings and how analyzing their creative process leads to a more general understanding of creativity. From this platform of psychological insight, we move to how creativity can perform a key role in the professional sphere. The discipline selected as reference for this chapter is psychoanalysis. This is a very specific option in terms of disciplinary orientation: a more superficial, more action-result-oriented theoretical approach might have been more palatable to a wider audience of business readers. This would perhaps have been more in line with contemporary practices of business coaching. Here however I try to go deeper, as part of a general ambition to explore alternative approaches to recent business practices, including people management.

In the last decade, the attention for applications of this specific therapeutic school has been decreasing, as the organizational setting of corporations demanded consulting services in talent coaching that would mostly enable faster performance of transactions. Compared with such recent training techniques, psychoanalysis is slow, deep, and richly textured. These qualities certainly match the overall vision behind this book. On this path, we encounter the founding fathers of this branch of human sciences, Kris, Freud, and Jung, in order to build up concepts and ideas on the specific study of the arts and of the mind of the artist. We then widen our investigations to the more general field of human talent and skills. Here, I introduce a number of very practical, very actionable techniques for creativity training. We review the ideas by advertising legend, James Webb Young, and the practices of an innovative design boutique agency in New York – Pompei AD. Chapter 3 rounds off with a number of specific examples and applications of the psychoanalysis of creativity within the context of professional organizations, as based on the important theories of Canadian academic Elliott Jaques. The case study in Chapter 3 tells the story of Franz X. Messerschmidt, a sublime sculptor of the eighteenth century and the protagonist of one of the most insightful case histories of classic psychoanalysis.

Chapter 4 ("Manifestos, propaganda, and the art of marketing") moves from psychology to brand communications-related "findings." We focus on the propaganda techniques that enabled the launch and creation of a major modern art movement – Futurism. We investigate the communication and collective dimensions of artists' work, through the lenses of the modern art movement. We describe the value of violent conceptual clashes, of punchy storytelling, and of daring campaigns. Most importantly, we (re)discover the crucial role of vision. The specific format of manifestos is introduced from its historical roots, having been illustrated by our very own opening *Golden Crossroads* manifesto. Chapter 4 does not limit its scope to just discovering early 1900s modern art techniques: instead, it presents practical ideas by expanding the horizon to selected contemporary agents that tap into culture to make a difference for their marketing customers. It offers three practical examples of how taking culture on board at the level of marketing planning is feasible. It concludes by observing how a classic museum can become a marketing territory of experimentation by daring to mix tradition and provocation – with both international and local success.

In these chapters, we encounter great names of the past and the present: F. T. Marinetti, founder of Futurism and a protagonist of European art, and his pupil Fortunato Depero, who exported Futurism to New York and into advertising and design unlike anyone else. The case study in Chapter 4 examines the controversial history and the charismatic protagonists of this unique movement, which celebrated its centenary in 2009. The chapter also looks at a contemporary master, Damien Hirst, creator of the most expensive work of art ever, and his controversial 2008 gig at the Rijksmuseum (Amsterdam), one of the classic museums in the world. It reviews art foundations and consulting firms such as Mona Lisa (Paris), Thonik (Amsterdam), ISISUF (Milan), 515 (Turin) and Caramundo (Rio de Janeiro). These firms are there, for the benefit of business leaders and brand marketers, to make change happen by tapping into the extraordinary power of creativity and fine arts.

We discover and discuss how artists think, how they work, and how their work can be beneficial to brand marketers and corporate managers. We do this through our next ten "findings:"

- Finding 6: Understanding the mind of the artist.

- Finding 7: Understanding what human creativity is about.

- Finding 8: Understanding the actionable power of creativity.

- Finding 9: Understanding how creativity can be nurtured by company training.

- Finding 10: Understanding how creativity works in the workplace.

- Finding 11: Create conceptual platforms to achieve maximum impact.

- Finding 12: Create a truly encompassing vision.

- Finding 13: Create bold brand narratives.

- Finding 14: Create your brand by anticipating cultural trends.

- Finding 15: Create memorable encounters of tradition and provocation.

Part II offers encouragement to business readers to rethink their practices and their ideologies of how work works. General managers will be stimulated to look at their staff as powerful agents of change, thanks to their creative skills; brand marketers will be presented with a number of important concepts from both the past and the present of fine arts, especially in terms of vision forming, proposition launch, and propaganda techniques. On both topics, a set of practical ideas offers a platform to act on what is learnt, with the specific agenda to make these ideas and concepts as easy as possible in terms of their deployment in everyday business practices.

CHAPTER 3

Psychology of the arts: how it works for work

This chapter focuses on the ultimate engine of the arts – creativity. Creativity is classified in different terms according to different cultures. Some Anglo-Saxon definitions see it as a skill set that can be nurtured through training. Another viewpoint sees creativity as a talent, and purely individual: you either have it, or you don't. Of course, creativity is always influenced by context: in the age of co-creation and co-design, we cannot subscribe any more to the romantic notion of creativity being the exclusive talent of exceptional geniuses. It is now established that creativity is supported by a number of combined personal traits – those that enable a constant reorganization of people's individual spiritual and/or public life. However it is looked upon, creativity is vital to business as much as it is for fine arts, today more than ever. It will take a "revolution of the mind" to innovate the way we work, think, and define our own business reality, in order to change as is needed to face the challenges created by the financial crisis. Creativity can help us all out.

This chapter explores what creativity is all about, and how it works for artists as well as business people.

The business question behind this chapter

What is creativity, and how can it work in the business context?

An introduction to the findings in this chapter

Ultimately it is we who decide what is art and who is the artist. It is we as collective members of a society that, at each and every moment of our human history, sets the standards, the hierarchies, and the rules of the fine arts game. It is always important to be aware of the social construct of the notion of art. Relativism is one of the great assets of humanism. From qualitative schools of human sciences, we identified key findings on creativity. It is not our purpose here to offer an exhaustive theoretical picture of the various schools and theories of dynamic psychology. On the contrary, experts in the field will perhaps find this chapter incomplete, just vertically highlighting some pinnacles in the psychoanalysis of the arts and of creativity. The second part of the chapter moves on to different theories, in order to horizontally widen our review of the topic from one person to a group or an entire company context.

The start of this chapter could have been set at around 22:00 hours, on August 11, 1956, in Springs, New York, when an Oldsmobile convertible fatally crashed, and Jackson Pollock lost his life and became a legend. Or it could have started with an interview with Damien Hirst, Jeff Koons, or Takashi Murakami, or any other contemporary living artist, discussing how and why they work in the way they do. Our journey into the mind of the artist could also have started in the eighteenth century, in Bratislava (see the case study about F. X. Messerschmidt, a classic case of psychology of the arts).

The challenge to understand the creative faculty of artists in all its facets and dimensions is complex, universal, and timeless. We will immerse ourselves in the mind and the soul of the artist, in order to understand what their main source of imaginative power and social charisma is – namely, creativity – and how it works. We explore very different definitions, notions, and metaphors describing the creative process, their motivations, and what artists think and feel when they are producing their work, and why. Because it is highly relevant, we will see how the process of creativity and the psychological traits of artistic creation are universally valid – after all, artists are human – and how such applications can be employed in the everyday business context. This chapter delivers vital insights to business leaders and people managers to improve their understanding of how their staff *really* work, and how the standard practices and policies of HRM and training can be improved accordingly.

The chapter as usual is structured around five "findings:"

- understanding how the mind of the artist really works from the viewpoint of psychoanalysis, in order to understand the general process of creativity at an individual level

- understanding the metaphors of creativity in order to explain what creativity is about and extend our understanding from the rather academic realm of dynamic psychology to the everyday context

- understanding operational definitions, to frame the different levels of creative performance, and their potential output in terms of innovation power: a map to enable business leaders to understand what it is that creativity can do for and in their organizations

- understanding creativity in the creative industry, as in the approaches, and especially the practical training methods, that can be learnt from leading examples in advertising literature and in the design world of today

- understanding situations and cases of how the creative mind works in the practice of everyday business, with – for example – a review of the impact of the mid-life crisis on the creative process, and beyond.

From fine arts to business, this chapter enables the discovery of one of our most treasured capabilities: the skills and the talents to rethink the world, and the ways that make such talent actionable, in the best interest of our companies, our businesses, and our life in general.

Finding 6: Understanding the mind of the artist

Creativity through the lenses of psychoanalysis

In the last ten years or so, there has been a progressive change in the nature of applied psychology to the people management challenges of everyday business. A few decades ago, the ambition of psychology was to enter the mind of people, and to understand in depth their emotions, their symptoms, and their meaning. In the last ten years or so, consulting services in this field have reflected much more the leveling of corporate business practices towards minimum common denominators. The general tendency therefore has seemingly been to stay on the surface of the psyche and just skim the bare minimum that is required to enable better performance of interpersonal transactions, with the focus on efficiency and effectiveness. Coaching programs have not aimed at engaging in the deep understanding of dynamic psychology: instead they mostly provided operational toolkits to make people work faster to – superficially – deliver better results.

In this chapter, it is our aim to discuss a different approach to read the human mind – psychoanalysis. Although it has not been *en vogue* for 20 years, this discipline has the advantage of enabling a different viewpoint, and a penetrating look at the ever-changing universe of our minds. To professional business readers, this "finding" might seem academic, or even redundant: after all, this is not a book about Freud and company! Please allow me respectfully to disagree with those opinions: the fact that a business publisher, in the year 2009, published the next few pages should be an indicator that a major change is needed in our training programs, in our people management, and in the way we regard the subject of work in general. With the help of psychoanalysis, our main toolkit in this chapter, we will be able to start doing so.

When it comes to psychoanalysis of fine arts, the first question is: "Can we study artistic creativity by applying the same tools as we use to study psychosis, neurosis, or fetishism?" And even more pressing as a question: "Can we learn more about the 'rest of us' from an understanding of the mind of the artist?" These questions have been at the center of at least one or two lectures, essays, and explorations by each of the leading thinkers in the field of psychoanalysis. First, although not in chronological order, the art critic turned psychoanalyst Ernst Kris provided a major body of work on the topic. Kris studied a great number of cases of artists' behavior and art manifestations, both pathological and not. He analyzed how artists' techniques re-establish magic connections between images and objects. He dissected the key role played by the narrative formats of artists' biographies in social terms. To him, myth, tradition, audience expectation, and artists' experience and psychotic delirium are all from the same system of idea generation within humans. However Kris regarded art as independent of trauma, and made a very clear distinction between the notions of "artwork" and "psychotic symptom."

According to Kris, artists are individuals with the ability to manage flexibly their regression into the darkest zones of their individual psyche. They tap into their imagination by taking a journey back to inner sources below the surface, where most of us do not dare to penetrate. At the other end of their journeys, artists do not just generate representations of nature, they stage a recreation of the world, with the positive social benefit of an increase in their own self-esteem. Kris indicates that creativity is divided into two fundamental moments: inspiration, or the journey into the deeper soul, and the disciplined fine-tuning and skillful optimization into the final work. It is the inspirational phase that might become fatal to those with a pathological condition. Kris's work is probably one of the most complete psychoanalytical theories about art, creativity, and the working of our minds.

Of course, Kris did not work in a vacuum when he was researching and writing on these topics. Before him, Sigmund Freud, the man universally credited with the foundation of psychoanalytical sciences, explored a number of individual profiles and cases from the history of fine arts, including giants like Leonardo and Michelangelo. Psychoanalysis was born at the end of the 1800s, and it was originally designed to transfer to the study of the human mind the applicability of thermodynamics and of other scientific laws. It is therefore no surprise that Freud gave the very basis of human creativity the name "sublimation." Within this one word, the superior "sublime" of beauty and art meets the more mundane chemical process of transformation of a solid body into gas. This key word is a milestone of the study of the artist's mind. Freud provided his vision of how sexual drives of an erotic nature are the basis for the performance of socially respected and intellectually praised work, including aesthetic and scientific research. This is what sublimation is all about: the transformation of what Freud defines as libido energy into intellectual value, for the pursuit of social success. How does this process work?

Disappointed by the limits of his real world, and pushed by an ambition to attain honor, glory, and love, the artist projects into fictional characters, and builds a world of imagination, with the ultimate goal of achieving success in real life, at the end of the artistic process. Within a fine balance between self-protection and narcissism, the artist takes off from the everyday, and invests their whole self in their imaginary life. It might look like escapism: it is actually the opposite strategy, as it implies the risk of failure and it requires the ability to cope with the risks of nonconformism and individualism. For Freud, creativity, or maybe to put it better sublimation, is a highly individual process, with roots in the play of our childhood, and also in our deepest personal drives and sexuality just like any neurosis, psychosis, or psychological pathology. The latter opinion is a point of major divergence between Freud and a number of his colleagues, especially his former star pupil, Carl G. Jung.

Jung's main point of disagreement with Freud on the psychoanalysis of creativity is simple, yet not reconciled: he not only sees a clear divide between art and trauma, he actually denies Freud's vision that the same pathology-focused tools of psychoanalysis may be effective for the purpose of understanding the artist and their creative mind. To Jung, artworks are symbols in the highest meaning of this word, well above any symptoms. This is, of course, unless pathology is at work: in that case we should speak of just symptoms, and apply the tools of pathology treatment. But if it is art we are confronted with, then its manifestations, the artworks, do not emerge from the drives and desires of individuals, but from a much higher source. Jung calls this source the "collective subconscious" – that is where archetypes reside. Archetypes are the collective DNA of the human race, and the most powerful engines of our imagination and soul. Archetypes do not belong to just one of us, artist or poet – they are a collective encyclopedia of human symbols. They belong to the human race as a whole and – through those special individuals who produce art – they speak to all of us. Archetypes reach us through myth, religion, and art, hence the "sacred" nature of art. Within the creative process, the artist might be in control or not, and – as long as the artwork is not finalized – they might experience it as an autonomous complex in their mind and soul.

If we follow the lines of this theory, the artist seems to play a shamanic role of a pseudo-religious nature. It might be because of the centrality of such a superior function that the artist is socially allowed to deviate from the standard norms that are compulsory for everyone else. With the freedom to explore the most remote regions of our soul, an artist provides a collective voice for us all.

At the end of this sixth "finding," we have expanded our analysis in two parallel directions:

- We have a wider hypothesis about where creativity comes from, what it is, and how it works in the mind of the artists: this will be the theoretical basis for the rest of this chapter, where these lines of thinking will be developed into actionable insights for the business reader.

- We rediscovered a non-transaction-oriented theory of the way the mind works: this is in itself an important viewpoint to bear in mind.

Both assets will become precious in our future in terms of business leadership, with particular reference to the creative industry, as the next four "findings" demonstrate. To further support this vision, we move to a wider analysis of creativity, beyond the worlds of dynamic psychology and fine arts.

Finding 7: Understanding what human creativity is about

Multidisciplinary metaphors to scope the general field of analysis

After the exploration of just one approach – psychoanalysis – dissecting just one theoretical domain – artistic creativity – it is time to engage in a wider interdisciplinary study, widening the borders of methodologies. Our approach to developing this seventh "finding" lies in a number of descriptions. In the early 1990s, in researching for a book, the Italian human sciences expert Alberto Melucci and his team of scholars and experts investigated what qualified respondents associated with creativity. We will find out what the outcome of this exercise was. What appears below is how the nature of our creative faculty is multi-sided and universal. This compilation of descriptions of creativity offers a general orientation of how the topic can be expanded horizontally beyond the vertical insight of psychoanalysis. Here we shift from the "artist" to "the rest of us."

Regardless of different theoretical interpretations, all psychoanalysts have studied art and symptoms, and sometimes art in symptoms and the other way around. The relationship between pathology and artistic production is an important one. Finally, creativity is a faculty potentially in the hands of healthy and balanced individuals as much as it is in the hands of unhealthy individuals. And it is not only individuals – if we think of contemporary art forms such as happenings and jazz improvisation, the individual dimension of creativity is rapidly immersed into the collective and destructured context of relatively unorganized masses. Groups and teams work according to specific processes: here, creativity and its management are also important. Even the masses acting and reacting in disorder can be studied as creative. By extensively using social science methods to investigate the topic of creativity, Melucci and his team extracted a number of descriptions, questions, performance indicators, and more. Reviewing a selection of these is the next step in our journey towards a deeper understanding of the art of creation.

First, we start with a number of words resulting from the question: How can you describe creativity? The work of various scholars and selected authors in social sciences generates a short list of words evoked by the noun "creativity:"

- maternity
- seeding
- mosaic
- light
- erotic pleasure

- dance

- mystical experience.

It is interesting to note how the word "creativity" stimulates other words that are both feminine and masculine, and how it covers a number of elements that recall ancient religious experiences and mysticism on one hand, and physical pleasure and erotic sparkle on the other. We find in this list of words all of the psychoanalytical references previously mentioned, from libidic drive to shamanism. If we focus on creativity not in terms of analyzing its aesthetic outcome but as a process, it is possible to identify from the book based on Melucci's research two descriptions:

- peaks within a continuous experience, with an accent on continuity

- Newton's apple, or the sudden "bolt of lighting" unexpectedly bringing the creative solution to a challenge.

This is interesting, as it describes the creative process both in terms of its climax, or peak, and as a more structurally built framework, with a longer extension in time. The key question is: Should we define the creative act as just the "Eureka!" moment that unexpectedly happens in the shower, or should we consider the latter as part of the creative act that includes long hours of tedious exploration before the actual bolt of lightning? Ultimately, is creativity a largely unconscious process, as indicated – although with different nuances – by the likes of Kris, Freud, and Jung? Or should we perhaps subscribe to more rational viewpoints, ones that see creativity as the ordinary skill of every child in any playground in the world? Is creativity a talent that comes with birth, or a skill that can be learnt by anybody with diligent practice?

This latter difference is fundamental: Talent is innate and unique, whereas skills can be taught and learnt. To take an example: as described in the case study in this chapter, with sculptor F. X. Messerschmidt, should we consider as skills his abilities to finely shape metals, and as talent his own aesthetic vision that made him switch from late Baroque to neoclassicism, just before (or just because?) a psychotic disease kicked in at some point of his life? What is vital at this stage is not so much to define *the* ultimate answers, but to identify the widest possible field of options. After all, creativity, together with love, is one of the most mysterious and intriguing features of humans.

Within this "finding," we extended our understanding of the creative function within the human mind. It became clear that creativity is a rich palette of opportunities. It is useful, as a next step, to structure its processes in categories that can help business leaders to understand its impact on projects, organizations, and even company/industry paradigms. This is the purpose of "finding 8," presenting what creativity can do for creative industry business success.

Finding 8: Understanding the actionable power of creativity

A systematic exploration of applied creativity for business success

From the first two "findings" in this chapter, we can state that the notion of what artistic creation is, and what creativity in general is, cannot be captured by a limited, and limiting, definition. We should instead work on a notion of creativity as a "conceptual container." By this we mean a verbal structure comprising more elements, which frame creativity as the human ability to generate change and to pursue the "new." When it comes to the world of fine arts, for example, creativity is about artists and their ways of working, it is about the context in which they operate, and it is about the final deliverables of their work.

This "finding" presents a classification of the actual impact of creative power on reality. As loosely derived from psychology of the fine arts, a parallel with business is related to different options of creative break-through. We analyze creative talent at work through three distinct and different levels of fit with the existing company/industry paradigms. I formulated the classification below in my dissertation, written in 1994, as based on 1970s research work by Italian scholar Gabriele Calvi. In 2000, McKinsey authors Merhad Baghar, Stephen Cooley, and David White analyzed this topic from the viewpoint of industrial processes and business consulting in their book, *The Alchemy of Growth* (2000). Our focus here is on Calvi's work: it is time to discover the potential of creative thinking for our professional universe, beginning with its most elementary level:

Level one: new combinations of existing elements

This level of creative work pertains to the exercises of members of a certain school: it is a codified school with precise rules and a clear theoretical framework to follow. Here, we see individuals remixing content according to a defined set of elements and rules. Sometimes the outcome can be a masterpiece: entire streams of art are based on the strict interpretation of a certain canon.

Within the business context, this is the level of change in continuity. Work excels when it matches the key performance indicators constituting the standard reference for the overall company and mostly for an entire industry. Here, creative professionals apply the rules as they are, with skill and with fluidity of execution. Think of standard 30-second advertising commercial creation in an average agency.

Level two: a change of frameworks and structures

This level of creative work can be exemplified by the change of style from late Baroque, a rather over-decorated style, to early neoclassicism, a severe and almost minimal style, at the time of Messerschmidt (see the case study). Here, we see two different ways of doing sculptural portraits. However, it is always sculpture, and it is always figurative. Nevertheless, we have two different visions at hand, and we can speak of a different (stylistic) framework.

Within this context, we can position medium to large corporate programs aimed at finding new ways in the context of the current business model. This is the place for the development of initiatives such as "AlwaysOn" by Hewlett-Packard (HP), where (thanks to Rule Developing Experimentation (RDE) by Moskowitz Jacobs, Inc.) a virtual discussion with customers and especially prospective customers was channeled into the process and work of design, prototyping, and engineering in real time, connecting and compacting the entire company around experiments that "stood in" for real people. The business model stayed the same, but the specific injection of elegant algorithm-based online RDE experiments changed its reach, its impact, and ultimately augmented the value delivered to customers by HP teams, resulting in a de facto internal organizational revolution in ways of working.

Level three: a whole transformation

This third level of creative working results in a completely new viewpoint being defined by avant-garde leaders first and accepted by art audiences next. You can think here of the moment Jackson Pollock dripped his paint for the first time, breaking new ground in terms of both poetics and technique. Or you can think of Wassily Kandinsky, and his first abstract image. These are the moments when the entire course of a discipline, in this case fine arts, is suddenly different, and the frame overarching it is radically changed forever.

This is the territory of corporate mavericks or start-up entrepreneurs who simply redefine the world according to their vision, and manage to change it accordingly. We can think of Steve Jobs and his insight leading to the Apple vision, products, and retail presence, because this is the league and level we are addressing here. As much as creativity is a skill that can be taught, it becomes evident from the example of Apple how different factors play a key role in success at this level, including charisma, vision, communicative power, and sheer audacity. Although we cannot subscribe to a notion of creativity as the outcome of romantic genius, not all of these factors can be learnt: it ultimately all boils down to individual talent.

We are dealing here, in conceptual terms, with three very different ways to apply creativity and to experience its impact. Take a master at the first level such as Salvador Dali. He was a technically impeccable painter, whose subjects were very disturbing to conservative audiences, at least until the 1930s. Those who judge the quality of his creative output within the framework of figurative painting might appreciate his work as showing fine technique: after all, Dali used to refer to classic art. Of course, those who are in favor of, say, religious art and against surrealism, will acknowledge his technique but disapprove of his content. On the other hand, Dali will be inevitably perceived as a reactionary by those artists operating according to third-level parameters: for example, the pioneers of abstract expressionism such as Pollock, who moved beyond classic painting, or the protagonists of the American land art of the 1970s, who destructured art into nature and landscape.

Does this classification of creativity mirror an implied judgment on the value of these three levels? We should be clear: this is not the case. Of course, some might admire or prefer one level of creative disruption of the status quo over another. Some see a greater degree of greatness in the work of those who change paradigms and reinvent the bigger picture. In reality, within fine arts, it is quality and *only* quality that determines the real value of creative output. And in fine arts, quality is unfortunately to a large extent purely subjective, although it can be rationalized, verbalized, and explained.

Quality means that sometimes the excellence of those who work within given boundaries beats the ambitions of those who aim to reinvent the world. The ambition might well be there but they might not deliver an outstanding level of quality. Of course, this means failure.

Within art systems, it is the context that defines the actual quality of a creative exercise. Therefore we can fairly say that – beyond issues of art criticism or the art market – an objective and intrinsic evaluation of creativity just refers to the fit – or not – between the expectations of an audience and the work of an artist. Of course, beyond such parameters as differentiation or contextual fit, we can measure the outcome of creative exercises in various other ways, including the number of deliverables. In times of economic growth, "quantity" can be a standard, as the market voraciously demands new work to exploit commercially with fast turnaround. This is the case with contemporary Chinese art, for example, as we will see in later chapters.

We have already drawn a number of connecting lines between creativity as seen from the viewpoints of fine arts and business. The key question then is: How can we apply some of these findings to our everyday professional practices? The answer to this question refers to our next "finding," where we focus specifically on the worlds of advertising and design.

Finding 9: Understanding how creativity can be nurtured by company training

From techniques to generate ideas to ideas for organizational training

It is now time to use what we learnt investigating the artist's mind and the analysis of creativity by moving to a better understanding of the mind of the business leader or the corporate employee. The main assets offered by this "finding" are:

- Further fine tuning of the classification of theoretical levels of creative achievability, as derived from the history and literature of advertising. Brand marketers will therefore find additional examples and references to consolidate what was explored above.

- Actionable ideas on how creativity can be nurtured through classic literature and nonconventional forms of company training program. Here, business leaders will find examples from the creative industries of advertising and design.

A few decades ago, advertising was seen as one of the hottest areas in which to exercise creative talent. In reality advertising, just like art, enjoys the coexistence of great innovators, who break all the rules, and "creatives" who simply repeat the same formats. Yet some do so with elegance and skill. Think again of a later Dali oil painting: total cliché but extremely seductive to some viewers. In the earlier years of marketing communication, the heroes of the trade, the fathers of the profession, wrote some very objective and rational manuals on how to be successful in creating advertising. Rosser Reeves' *Reality in Advertising* (1961), a textbook prescribing the exact way to create advertising, is such a work. These books were aimed at creating rigid sets of rules for the execution of communication formats: the ultimate formula. The end result was the perpetuation of the same messaging structures, from the 1950s onwards, for entire categories, from analgesic medications to toiletries and food.

In the history of advertising, we can speak retrospectively of a constant tension – if not open warfare – between the innovators of the industry, and those who found their ultimate comfort in the dogma of the formula. It must be said that – unlike with art for art's sake – within the realm of advertising, innovation for innovation's sake does not necessarily work. Sales generated by adverts and their memorability can easily be tracked, and this shows how things tend to go wrong if there is an attempt to force the paradigm, without the necessary talent. Much better, then, to pursue the elegance of a different execution within existing formats.

In the vast archive of (mostly frankly useless) books about advertising creativity, how to be creative, and how to innovate, what is still considered the best by many is *A Technique for Producing Ideas,* a very short book by US advertising executive James Webb Young (1940). Young created a light manual, simply describing the few steps he experienced as the best approach to creating. His book was written for his students at the School of Business of the University of Chicago. Written in plain language and in a reader-friendly style, this manual demonstrates practical steps to generating ideas. New ideas, according to Young, are all about finding new combinations of existing elements. The research explorations underpinning the creative process are described as an accumulation of information: both specific, on the topic at hand, and generic, to fertilize the process. Then there is a kaleidoscopic remix of all accumulated information in the form of individual brainstorming – the abolition of any (self) censorship, free association and the recording of every possible new combination. It sounds like fun, but if you want to make it really work, it is hard work. Once you have stretched yourself and expanded the field of your associations, you simply have to relax and stimulate your imagination by doing something completely different, at best something that you love doing. Whether it is listening to music or doing physical exercise or making love, what is vital is that your mind is left working on its own. This is not time wasted, it is the time to let ideas emerge. It might seem odd, but this thin, elegant, inspiring book was written in 1940. We should consider James Webb Young and his excellent book as our key point of reference in terms of applied creativity to the business context. His suggestion is actionable, it is practical, and it is rooted in decades of successful advertising practice by its author.

After learning from one of the best advertising authors of the last decades, let's explore an example of creative training from the design world. This does not come from a book, it is taken from everyday business. Here, the devil is in the detail of standard business processes, even when we look at the very core of the creative industry. It all starts with a simple question: What is the common notion of talent training? Corporations identify their most promising talents, and invest resources in training them, so that these individuals can generate differentiating ideas and competitive performance results. This is the general idea, but what about the actual execution? The training formats for corporate staff on a growth path are largely the same: they all follow the MBA format. In essence, all this unique and promising talent receives more or less the same content, based on the MBA vision of the world. To use an analogy, it would be like sending Picasso, Dali, Pollock, and any other artist to learn exactly the same techniques, within exactly the same paradigm. What kind of truly creative edge can be derived from this standardized approach?

The good news is that there are alternatives to this MBA ideology of training. Let's learn from leading boutique architectural and design firm, Pompei AD of New York. Ron Pompei, CEO and chief creative director, is a former artist and a respected art collector, with a full understanding of the contemporary art sector. As enterprise leader, through the years, he systematically nurtured the power of fine arts by investing in a small gallery, C3, with a very strong research orientation. Although a separate business, the gallery was located next to the office space of his firm. The gallery is managed by an independent curator, and organizes shows, lectures, and networking events. This established a first reference point for Pompei AD staff in terms of creative inspiration. Think of it in terms of a permanent fitness room of the mind for the entire company, located just next door.

On an individual training level, Pompei AD adopted a very particular HRM policy. Beyond standard professional courses, each employee received a special bonus of US$2,000 or so per year, to be spent on activities of their own choice, no questions asked. This way, each member of the staff could enjoy the freedom to pursue their personal passions. Only one condition was given and one obligation existed: to prepare a public presentation for their colleagues and studio staff, and share their experiences gained while following their passions. By sharing experiences, the entire firm would have each member coaching the rest of the team about their vision of the world and their personal interest, and everybody could learn from each individual. What better approach is there for a business that believes in the individual talent and human potential of every single staff member?

Finding 10: Understanding how creativity works in the workplace

From preventing crisis to managing opportunities

Whether it is in advertising or design or consulting, if we are willing to use the power of creativity in our professional life, there is more to explore. To do so, we turn to Elliott Jaques, the Canadian author, who in the 1970s applied some specific principles from psychoanalytic research on creativity and fine arts to the world of business. Jaques used as a reference work by the Hungarian analyst and creative thinker, Melanie Klein. Klein established an original path by focussing her investigations on the feelings of aggression that children and babies experience toward their mothers. Within Klein's vision, each of us builds an emotional landscape of objects. As babies, we experience our mother's breast as a good object because it provides us with nutrition and comfort. However, when such an object is not available, we rapidly become frustrated, and develop hate towards the very same breast we used to love, experiencing it as a bad object. When we hate our bad object, we feel paranoid fear for ourselves. When we realize that our bad object is actually the same as the good object we depend on, we feel depressive anxiety and fear for others. At that moment, we strive for ways to mentally protect and repair the object we so much hated, because we know that it is the very same object we do depend on. Melanie Klein concluded that the content of these fantasies and the individual balance we achieve among these different internal responses are the ultimate basis for our entire psycho-emotional life, including creativity.

Within this theoretical context, working as the director of the Institute of Organization and Social Studies and of the Health Services Organization Research Unit at Brunel University in the United Kingdom, Elliott Jaques

(1970) classified the nature of work into three different typologies of creativity:

- Creative work that uses the pure symbolic level, without a framework of reference. Think of groundbreaking artists or highly innovative creative directors in the ad industry.

- At the other end of the spectrum, Jaques sees bureaucratic work, where employees act according to a simple frame of reference. Here, the main reference is the fit between employers' assignments and evaluation frameworks.

- At an intermediate level, Jaques positions the work of managers, who have a strong reference framework but also the facility to act in a creative fashion. Often managers juggle the limits of their assignments, sometimes actually determining them.

Mind you, this is an external description level only. Within our minds, each of us, even the most boring administrative clerk, experiences some level of symbolic creation, because according to Klein each of us has to deal with their own internal objects. By doing so, we inevitably activate symbolic projections onto the outside world. Jaques' conclusion is that work, as in our everyday work, is largely influenced by our need for reparation of our Kleinian objects. When things work fine, then work is a positive flow of healthy creation of symbols. There, we can speak of harmonious relationships between company, social dimensions of work, and the interior life of employees and managers. Such harmony translates into flow and engagement, and the workplace is then inhabited by humans who perform their tasks with a great amount of sublimated energy and with a strong sense of existential purpose. To some extent, in this situation the mindset of workers and managers would not differ from the dynamics of an artist at work in their studio. Things greatly change when the workplace generates discord. The individual outcome, when things do not work, is always negative, although there are several different forms:

- *magic defense*: magic omnipotence, grand plans, lack of accuracy in the details, early delivery of incomplete work

- *obsessive defense*: excess of details, lack of imagination, lack of decision, delayed delivery

- *schizo-paranoid regression*: inhibition of the thinking process, destruction of symbols, confused organization of internal work, an external impression of stupidity.

Do not get too impressed by the vocabulary: the definitions above are related to technical descriptors of common pathologies in dynamic psychology. Nevertheless, what is evident is the fact that the interaction between mind and work can deliver, in the context of office space, equivalent symptoms to the ones experienced by pathologically affected artists. It is vital to understand the symbolic working of the mind, and use such knowledge to anticipate a potential crisis.

If we go back to the creative industry, and look at creative directors, design directors, and other such business leaders, we can extrapolate from the work of Jaques a specific study of one particularly acute phase in their lives, as professionals and as humans. A typical turning point, especially in the professional life of creative leaders, is the mid-life crisis and its impact. This is the delicate step where once hot young talents become calm, professional, mid-life mentors and personal coaches for large teams of talented but younger members. Jaques identified two possible outcomes of this very delicate professional and personal moment:

- In the case of a positive resolution over time of the original interior conflicts among Kleinian objects, the creative or business leader will reach the time of their mid-life crisis ready to accept their destructive drives as much as the inevitability of their own death. In such a case, Jaques states, the talented creator is now ready to become a mentor for a new generation of professionals in their own field, and to pass on the torch of creativity and excellence. The development will then be equivalent to the evolution of an artist working as Mozart did, all intuition and genius, into the patient and the methodic approach of a mature sculptor, capable of expressing their talent in different stages, from idea to refinement.

- On the other hand, if the mind of the professional develops towards unresolved objectual conflicts, they will see hate and envy prevail. This will determine a highly disruptive mid-life crisis, with the progressive freezing of all processes of sublimation, and the end of all creative faculties.

The latter is clearly a pathological situation, with serious consequences for both the individual and their team. You might have recognized in Jaques' shortlist of work-related pathologies some of those typical biographies of romantic geniuses or the deviant artists of any age. You might as well recognize some of the behaviors displayed in your working environment, and at some point affecting (sometimes unexpectedly) some of your staff or peers or managers. The point is simple: these symptomatic behaviors might well be signs that these professional people are not well, at least from a spiritual and psychological viewpoint. If you are a business leader or people manager, it is important not to address such unrest purely from the functional or operational management perspective, but to think at the deeper level where emotions and creativity work. Find new ways to exercise your authority to change the environment around your people, to make a difference for them, by helping them in nurturing and managing their creative talents in a better way.

Wrapping up: from finer creativity to a better quality of life

Nurturing talent and putting creativity at the heart of business practice is obviously not impossible. It is however a characteristic trait of boom periods, times when faith in the future and exploration of opportunities are maximized. Think of the late 1990s. A new generation of talents dreamt for a few years of redesigning the way we live. The hopes of the IT revolution – a cultural revolution that produced the Internet dotcom boom in the early 2000s – were to liberate the minds of each of us. Within the corporate and general business context, some might say that this mostly resulted in the replacement of staff by means of e-suites and applications, with programs generally aimed at just cutting costs. In the end, looking at contemporary corporations, the side-benefit (lower costs) became the main objective, and what had been the main promise (empowered individuals and a thriving organizational culture) never really happened.

Think of the 1960s, and the age of the creative revolution in advertising. From the creative edge of those days, through the decades major advertising networks turned their business into what mostly now looks like corporate business, with the same open space environments, the same impersonal HRM policies, and the same sterile focuses as any corporate bureaucracy.

The unrest and the psychological pain of staff are particularly evident in times of downturn and restructuring, with the progressive spiraling of organizations into a cost-control stream first, and into management by fear in the end. As an example: creative people, including designers, scientists, and art directors, tend to have a very specific sense of space and of their own territory. Halving desk space is part of the classic recipe of cost management. By reducing desk space, the efficiency of building usage increases, and the costs of building management are reduced. As a final outcome of desk reduction policies, creative talent is frustrated and irritated. As a result of these policies, the best talent normally starts its exodus to other employers, for example universities or academic research. The rest stay and suffer, with a negative impact on the quality of their work. What benefit can come in the long term from these account-only-driven policies? No wonder that smaller boutique agencies structurally took the lead over large incorporated advertising firms in terms of quality of output.

The German sociologist Joseph Schumpeter made the notion of "creative destruction" well known. In the interpretation of MBA-driven managers, the term is an oxymoron, especially when you look at how managers in recent years have focused on the benefits of downsizing as one of its presumed best features. Beyond Schumpeter's acute analysis and its subsequent vulgarization by advocates of the hard-edged capitalism of recent years, the key issue is that the 2008 downturn proved one thing: a system geared towards numbers and profits failed to deliver universal and sustainable well-being. Why not learn from the true power of the creativity of artists? Why not redesign the way we work on the basis of a different principle, to be called "creative creation"? The bottom line is that the business world could use the creative understanding of the fine arts as a mirror, to look for the next and the new. Therefore, it seems appropriate to conclude this exploration of the mind of the artist by looking at one of the consulting firms that work with this objective, using art as the reference.

Enter Mona Lisa of Paris (created and led by Valerie Bobo), the innovative consulting firm that connects the power of fine arts to business process and innovation. Let's examine a few highlights of the benefits businesses can obtain from such an approach:

- the key relevance of establishing a vision of the world, and of pursuing the implementation work according to a visionary leadership principle

- the key importance of feeling the environment around you and the soul within you, in order to let them resonate

- the aptitude to take risks, to explore the unknown and to experiment beyond what is the established status quo

- a drive towards sensorial communication codes, where abstract ideas are made tangible and have an impact

- a profound respect for intuition as *the* way to capture the essence of situations and the thrill of the new.

This is what fine arts bring to Bobo's customers, straight to the heart of their corporate processes. At the same time, this is the kind of language that you might find in the vision and mission statements of large corporations and creative industry firms, of high-tech research labs and idea-driven industries. Nevertheless, most corporate structures at their very heart are mostly inspired by an opposite focus: number-crunching. Mona Lisa offers its customers the opportunity to regain a sense of humanistic dreams, turned into actionable corporate assets. Once the way back to these core values and ideas is found, it helps corporate executives to inject new thinking within the very heart of their companies – the business and management processes, in a practical and actionable way.

To recap, initially we embarked on a journey through the mind of the artist. By discussing some ideas of the founding fathers of psychoanalysis, we discovered how creative work requires a deep symbolic regression into the deepest territories of the psyche, at the risk of entering pathological zones. We then moved on to the business world, and here we found some important advertising literature of the 1900s. We went through the rational, yet almost poetic, approach to defining a technique to be creative, as described by James Webb Young in 1940. We concluded our journey with a few direct applications of psychoanalysis to the context of professional environments. To summarize, the essence of this chapter is:

- Artistic work is a powerful way to reconnect the deepest drives and emotions within the soul of a human being to the professional context.

- The creative mind of the artist can be analyzed at the level of dynamic psychology and other human sciences, with the findings of such explorations being extremely helpful in understanding universal mechanisms of how we think, feel, and work.

- In times when new ideologies will have to emerge in order to rethink our world, the world of fine arts and the psychology of artists offer valuable insights for the rest of us, with particular focus on people management and organizational training.

Let's think for a moment of our possible futures. What if contemporary idea-driven, creativity-propelled organizations could really understand the symbolic, deeper world of the people who work for them every day? What if they really put such people at their core? What if they could shape their governance according to the principles listed above?

Between inspiration and pathology: Birth of a classic psychoanalysis case history

It is the year 1781, and we are in Pressburg in central Europe, today known as Bratislava. We follow the German writer Friedrich Nicolai who is visiting an old master. This old master is a man who lives on the unappealing side of town, next to the Jewish museum. We feel a shiver as we enter the studio, and we certainly are not at our ease: this is an artist and a former academic who comments ungenerously on his peers, or – as he calls them – his persecutors: academy professors, court artists, and at some point, the whole of Germany. He has been convinced since his youth that the entire country hates him because of his talent.

We turn to his work, modeled by his sophisticated hand. It responds to the most advanced research of his times: physiognomy. The expressions on his sculptures, which are all of men's heads, are cold, frozen in a fixed set of grimaces. Nevertheless, the talent of the old master can be seen in each of them. By the end of his life, the old genius who lives on the outskirts of society will have produced nearly 70 pieces. Still famous decades afterwards, each of them is part of a coherent oeuvre.

The old master is Franz Xavier Messerschmidt. Conforming to the stereotypes of artist's biographies, Messerschmidt came from a modest German family, and reached a peak of fortune and excellence with important court commissions for the empress, Maria Theresa of Austria, and the princess of Savoy. He also became one of the most exquisite masters of the late Baroque style in the Alpine region. Since 1769, he had moved beyond this style, venturing into a rigorous neoclassical direction, built upon his fruitful trips to Rome. With such credentials, his past roles at the Academy of Vienna, and maturity beyond his years, he seemed to have a more than promising career ahead of him. His destiny was to instead fall progressively under the spell of his psychosis, and end up being expelled from the Academy in 1774. He moved to Munich first, and ended up in Bratislava where he shut himself away from the rest of the world.

Despite this, he still managed to use his sculpting talent according to the style and the trends of his age. Where did the creative inspiration for the series of 69 heads by Franz X. Messerschmidt come from? At a basic level, each of them was based on his own image as captured in a mirror. More significantly, each head captured the ghosts and deeper fears of a mind in need of help and relief. For the benefit of Nicolai who was visiting him, Messerschmidt explained his ways of working. He had to fight constantly against what he called "the demons of perfect proportions." He had constantly to reassert what he perceived as "the perfection of nature." He struggled especially with two exceptional character heads, the most deformed and least human ones. These two heads presented a strong elastic deformation of the face, and a stretched out tongue reshaped into what appeared to be a bird's beak. These were the ones he would never dare to recast, because these were the two heads that represented the visual essence of those demons that appeared to him at night.

The writer, Nicolai, captured the essence of these rare moments in his travel diary. More than a century later, these notes became the source for Ernst Kris, the art historian turned psychoanalyst, to diagnose Messerschmidt's psychosis, and to write one of his groundbreaking essays on the psychology of the arts.

CHAPTER 4

Manifestos, propaganda, and the art of marketing

Far too often these days, brand communications happen in an artificial environment and vision is simply not part of the mix any more. This is the case when the marketing of products is left to simple media techniques and conventional formats. On the other hand, the world of fine arts is a rich source of examples where artists and movements established themselves thanks to superior communication approaches. Some of the most talented artists in the contemporary art sector work with a clear marketing planning technique, from Takashi Murakami to Damien Hirst. Of course, this is not new: old masters such as Rembrandt had their own "branding policies." According to some, fine arts is about finding an inner vision and selling it to the world. We might say that entrepreneurship and marketing at their very best are actually the same. It is time to find out what fine arts can share with brand marketing professionals, and other practitioners operating in this field.

The business question behind this chapter

What can the world of fine arts teach the brand marketers of today and tomorrow?

An introduction to the findings in this chapter

In the years ahead, the ability to create a true brand will be the difference between success and survival. But what is a true brand? True brands are not simply economic assets, they are cultural agents. They go well beyond the sphere of marketing and branding, to establish a deep bond of love with people, sometimes across generations. True brands are loved and sometimes adored because they stand for something that people care about. At their very best, they stand for the cultural DNA of a generation. Think of Nike in the 1990s, before the child labor exploitation scandals, or Apple in the 2000s, and how their slogans – "Just do it", "Think different" – embodied the spirit of the age, above and beyond retail and pricing. Through the years, the ability to create true, authentic brands became blurred within the repetitive technical cycles of brand management and its automatisms. Branding is magic, and magic cannot happen by means of repeatable mechanics.

Modern art offers major insights on brand marketing to be treasured, such as the experience of the Italian propaganda masters of Futurism, the Italian art movement that had a major impact in the first half of the twentieth century. This was an art movement rebelling against the past, warring between tradition and the future, and taking an aggressive posture about art and life. Futurism reminds us how to launch a proposition – be it an art movement or a commercial brand, by extrapolation – with minimal means, or no means at all. This chapter also offers us the opportunity to reflect on how a brand should aspire to a level of ambition that goes beyond category management. How about rebuilding a world (or even a better universe) around people's lives? Mind you, this strategic drive to rethink the brand as a universal vision does not translate into the "operational" mandate to design to a specification the whole environment where people live. On the contrary: universal vision has to meet co-creative design.

We focus here on the dimension of vision and on the dynamics of launching a proposition. Brand marketers will be given food for thought to rethink brands as ideological – and perhaps even political – platforms. Here the format of the "manifesto," a strategic communication tool very dear to the leaders of Futurism and also adopted by this book, will offer the opportunity to regain a sense of the relevance of strong storytelling behind branding.

Considering today, two "findings" provide contemporary references for how to use these principles. In particular, to take a recent case, the story of Damien Hirst and his 2008 show at the Rijksmuseum in Amsterdam demonstrates how the clash between the traditional and the provocative can deliver exceptional marketing results. In parallel, the introduction of a few examples of art institutions and consulting firms offers an indication of how marketers can tap into the vibrant jolt of culture, in the same line of thinking as *futuristi* did.

The main benefit from these "findings" could perhaps be to recapture what was once the true talent of the grand entrepreneurial leaders and brand marketers of the past: to define their vision of the world, and translate it into a far-reaching proposition. These are the "findings" we develop in this chapter:

- *Launching a vision against all the odds*: How Futurism was created from scratch, and what brand marketers can learn from it.

- *Reconstructing a universe around your brand*: How a vision should lead the world to think beyond simple products.

- *The art of the manifesto as brand platform engine*: How a fine arts format can be the most powerful tool to channel a brand marketing vision.

- *Connecting fine arts and culture to marketing creation*: How selected foundations and agencies tap into the power of culture to deliver outstanding creative solutions for their customers.

- *Creating dialog between the classic and the contemporary*: How daring to be different makes for a success story – even in the most conservative museum setting, with the result of being dubbed a national marketing "guru" for those cultural and museum marketers who dare to dare.

Our case study discusses the history of Futurism, while our key focus in this chapter is to employ history in order to inspire change in the near future. This is very possible, as the show by Hirst in Amsterdam proved. In 1909 Marinetti aimed at destroying museums to jump-start the next wave of Italian and universal art, and in 2008, Hirst brought the impact of contemporary provocations in the classic setting of a traditional museum to a whole new level. As the Italian artists who created Futurism were focused on the future, our journey in this chapter is clearly focused on the future also.

Finding 11: Create conceptual platforms to achieve maximum impact

Relearning the art of launching a brand

Futurism is an art movement that tapped into the spirit of its times, with energetic vigor and violent passion like few other movements have done. Such connection between the essence of culture at a given moment in time and an agent in such a culture, be it an artist, a movement, a brand, is a vital asset for marketing success. Art scholar Marjorie Perloff crafted a memorable definition: "The Futurist moment" which is also the title of her book (1986) about Russian, French, and English protagonists in the movement. The Futurist moment is the perfect alignment between an idea, an art movement or, by extrapolation, a brand, and the spirit of its years. It is a climax at the level of symbols and signs. This is what 1977 was to the punk movement in popular culture. If we extend into the realms of design, it is the flourishing of Italian fashion and luxury brands such as Armani, Moschino, and Versace in the 1980s, or 1990s Dutch design, or the early 2000s dotcom lifestyle revolution. We might say that the "Futurist moment" is the Nirvana of any marketer: that exquisite moment when the brand seems to be perfectly positioned in the world, from each and every viewpoint. This is the phase of the maximum energy of the appearance of a cultural trend, and the moment of its greatest impact on society.

This "finding" is a great tale of how a powerful fine arts brand was created against all the odds, and without any major assets to depend on other than the exceptional talent of a half-dozen individuals. How did Futurism achieve its very own Futurist moment? Talent, boldness, and faith. However, as far this book is concerned: the answer is propaganda, or as we would call it today, "brand marketing." Futurism founder and patron, F. T. Marinetti, never doubted: image was everything, fame was the minimum goal, and bad press a great way to be noticed. He never feared negative feedback from audiences and mass media. He saw scandal as an asset, and this, retrospectively, is another element in our fascination with Futurism. Some of these techniques reached the contemporary business marketing world in the form of entertainment publicity (for example, the Sex Pistols punk rock group publicity campaigns in the late 1970s and early 1980s).

What were the communication strategies staged by Marinetti and his small group of artists and poets to achieve such a result? To Marinetti, the choice of media was crucial: the ability to write statements of great impact would not have been enough for an art movement in early 1900s Italy. This was because any statement coming from Italy would have been classified as the product of a provincial, marginal artistic scene. So Marinetti chose to launch the *Futurist Manifesto* on the front page of *Le Figaro*, at the time the leading newspaper in Paris, the capital of the art world. From here, an aggressive strategy was necessary for the small group to be noticed, and ultimately thrive. Crucial to the popularity of the movement was the ability to spread its message with the techniques of cabaret and theatre: the staging of highly disruptive events before the opening of Futurist evenings, such as the selling of the same theatre seat to ten different people, played a great role. This gimmick fueled tension and started arguments even before the start of the always controversial show, guaranteeing the highly visible negative publicity that Marinetti was seeking. Another example is in 1910 when the Futurists dropped leaflets from the top of St Mark's Cathedral in Venice which demanded the "creative destruction" of Venice and then its reconstruction as a contemporary metropolis.

Across and beyond their communication strategies, impressive elements in a general review of the movement were its ability to make a strong point of its vision in spite of its limitations, and its collective talent to generate ideas and regenerate itself under different, sometimes difficult circumstances:

- One example of the first "ability," the talent to work in spite of technical limitations, is how the group of founding fathers of this movement managed to get exhibited internationally with paintings that spanned various schools and genres, without marking out a real new territory. Cut across divisionism, cubism and various experiments to render speed and abstraction, Futurist paintings did not offer the aesthetic cohesion of other movements. In spite of these limitations, the vision of the movement was blessed by being shown respect in France, England, Russia, the United States, and all countries that mattered in modern art – and this for a century after its inception!

- The movement started with a hard core of five protagonists: Marinetti, Boccioni, Balla, Carrà, and Severini, all painters and sculptors, and then extended itself in various waves. New protagonists emerged, such as Fortunato Depero, with the ability to move the original visions and ambitions into new areas, such as graphic design and interiors, and in new cities, such as New York.

The lessons from this "finding" for today's brand makers are:

- *Concentrate your media focus on the most credible, most impacting platform*: The rest will come by viral diffusion.

- *Dare to be bold, to speak out and to create controversy*: From Benetton to Diesel, this is a winning strategy for lifestyle brands.

- *Do not hesitate to launch regardless of your own limits and shortcomings*: If you believe in yourself, the world will likely follow you, and you will build up on your powerful vision by inventing your own future as time unfolds.

In the old times of brand marketing pioneers, there was almost a sense of discovery and wonder: brands were not scientifically engineered, they were grown from scratch, and made into powerful assets, sometimes against all odds. The glossy MBA tooling that came with the last decades has provided a powerful array of enablers, and it has increased the sense that everything in branding is about ultimate perfection. This is sometimes translated into a high degree of rigidity by the makers of brands: quite a paradox for minds whose aim should be the pragmatism of constant creation. It is time to demand more from brands and their creators. Futurism can offer more inspiring ways of how to do so, starting with the level of brand vision.

Finding 12: Create a truly encompassing vision

Envisioning an (urban) universe around your brand

The real strength of Futurism was not necessarily any single execution of its poetic vision but the overall scope of its reach, and its ability to articulate its vision in the body and soul of its times. Of course, "words in freedom," Marinetti's own household poetical technique, and a number of 1910s paintings or sculptures did represent an outstanding level of aesthetic achievement. However, when the vital energy seemed to run out of the movement, its disciples turned the table of conventions around, and innovated in new fields of "lower culture" rather than "high art" – advertising, radio drama, and interior design. This ability to look at the world and always catch a magic glimpse of its cultural spark is what enabled Futurism to remain relevant until the 1940s, and one of the main reasons why Futurism still matters to us so much today.

From fashion to carpet design, from cuisine to lust, every single feature of what we might somehow vaguely identify today as a "lifestyle proposition" appeared on the horizon of the Futurists. Marinetti dedicated an entire book, *Come si seducono le donne*, to the art of seduction, and Valentine de Saint Point wrote a manifesto to celebrate lust and erotic desire. As one of his groundbreaking manifestos, Marinetti launched the "cucina futurista," a kaleidoscope of recipes aimed at revolutionizing the consumption of tomorrow's gourmets, beginning with the abolition of pasta and the chemical creation of whole new sets of flavors. The list could continue: architecture, politics, and more.

There are not many aspects of life on which Futurism did not express a point of view, an opinion, or an idea. The ambitions of these artists were of universal scope: for example, "The Futurist reconstruction of the universe" is the title of a 1915 manifesto by Giacomo Balla and Fortunato Depero. This manifesto came in the form of a short leaflet. It included references to Boccioni's visual visions, to Marinetti's words in freedom poetic technique, and to the technical requirements of a Futurist toy for tomorrow's children. As usual, it was a rather encompassing, somehow inconsistent, yet powerful mix of topics. The challenge behind this specific manifesto is how art can dissolve itself into life. The main stage of the Futurist vision of an encompassing ideology of life was of course the contemporary metropolis.

By definition, new lifestyles have been, are and will be urban in their embodiment, and Futurism was *the* urban movement par excellence. Given their vision, it would be natural to expect Marinetti, Boccioni, and all the other founding fathers of Futurism to feature regularly in a city such as New York. After all, if you are serious about being a leader in, say, architecture and design these days, how could you claim you had never visited Shanghai or Tokyo? In the 1910s and 1920s fine arts field, New York was not what New York is today. Its metropolitan leadership status was acquired after 1945, with abstract expressionism. The roles of political and cultural capitals of the world before the war were more suited to London and Paris. Perhaps the contemporary equivalent of what New York was then, in the first half of the twentieth century, is the Tokyo of the last ten years, or the Shanghai of the last five years, or today's emerging cities such as Istanbul and Dubai. It is the place to be in order to spot the emerging trends. It is one of the world-class booming cities, with a dynamic population, a winning mix of immigration and growth, and a sense that the future, whatever the future will be, will start from there. Antonio Sant'Elia, the visionary architect of the Italian movement who died in the First World War, saw some printed images of Manhattan. On that basis, he imagined his "città futurista," with skyscrapers and train station terminals and hanging gardens. Sant'Elia never managed actually to build anything. However, he remained a major source of inspiration for architects worldwide for the next century. Such is the power of vision, the energy of dreams.

Marinetti, Carrà, and the others associated with the movement traveled to Paris and London, to revolutionary Russia and to the countries at war in the 1930s and 1940s. But the only one among them to spend time in New York was Fortunato Depero, the author of the manifesto "The Futurist reconstruction of the universe." Futurism enjoyed one of its high points in Depero's American experience, and the work he performed in his Chelsea studio. Here he produced rich and important projects, for example as the art director of several 1930s covers for *Vogue, Harper's Bazaar,* and the *New Yorker*, and as creator of advertising work of the greatest quality, in line with what he did before and afterwards in Italy. We might naturally expect an artist with such vision and a graphic designer with such talent to find himself in America. Nevertheless, perhaps also because of the difficulties of the Great Depression, Depero never became naturalized. He did use his own Italian PR techniques to attract customers, such as preparing and displaying delicious pasta for the visitors of his shows. He achieved some success with his talents, but in the end he was to be one of those migrants who went back home, in his case to the town of Rovereto, near the Dolomites.

In his home town Depero built his "atelier" into his own museum. In the heart of Alpine provincial life, he created a Futuristic world of his own, designed according to the latest trends and taste. He founded an enterprise that recalls Andy Warhol's famous 1960s Factory, at least in some of its guiding principles. The result was a sort of creative laboratory established to produce tapestries, furniture (what would be called today DesignArt or autonomous design), and advertising. This museum was restored and reopened amidst much media coverage 100 years after the founding of Futurism. We can conclude that Depero's work has stood the test of time, with his highly versatile, truly multidisciplinary portfolio of applied arts.

Shifting from his 1930s to our 2000s, what are the lessons to be learnt from Depero's experience, for the benefit of today's brand marketers and business leaders?

- *Extend your vision from just a category to the entire universe:* You might not want to redesign people's everyday lives from scratch. However it is crucial to think in holistic, encompassing terms.

- *Dare to be where today's and tomorrow's trends are in the process of being formed:* As Depero did in 1928 and Akio Morita, founder of Sony, did in the 1950s in moving to New York, it is crucial to physically explore, in person and in depth, key countries at the moment they open as new markets.

Depero's days in New York City left lasting marks. Perhaps inspired by his American experience, he published the 1932 "Manifesto dell'arte pubblicitaria futurista," a vibrant document that anticipates some of the post-Second World War theories and themes concerning the relationship of advertising and fine arts. What made the specific medium of the manifesto so special to the protagonists of Futurism? And most importantly, what can marketers learn from such techniques?

Finding 13: Create bold brand narratives

Your brand is just a story: design it to be truly memorable

If a brand is all about a clear proposition supported by a great story, then a manifesto is about the DNA of a vision, communicated with precision and effectiveness. Of course, the most notable manifesto in modern history is Karl Marx and Friedrich Engels' 1848 *Manifesto of the Communist Party*. You might agree or disagree with the ideas, you might like or dislike the vision; however: nobody can deny the effectiveness of such a manifesto as a call to action, since it actually changed the world for decades. Prior to that, manifestos were more formal documents, without any deep communicative impact. The *Manifesto of the Communist Party*, from a literary and propaganda viewpoint, delivered a great opening paragraph and practical statements. That was a lesson that artists and writers who followed did not forget.

Marinetti's 1909 manifesto created Futurism. If this is the power of a simple literary construct, why do corporations and advertising agencies not work with manifestos? These days, each company and enterprise has its own brand platform, brand book, and brand DNA PowerPoint presentations. In the last 10 to 15 years, a tsunami of books, guru conference speeches, consulting hours and so on have resulted in a proliferation of brand strategies, brand design, and similar corporate projects. Needless to say, and as I have personally verified during conferences and meetings with various *Fortune 500* blue-chip companies, all these visual and verbal articulations of brands align around a few themes, such as "working together in teams," "servicing our customers with extra value," and "respecting people and the planet," mostly matched by standard images of corporate clerks smiling in their button-down shirts, all produced in the most neutral and politically correct mix. What is the differentiating value resulting from the millions and millions of dollars spent on corporate branding across the last decades? Not much, to be sympathetic, and just noise and undifferentiated mediocrity, to be critical. I recall a conference of the Design Management Institute where competing world-class *Fortune 500* brands presented their corporate identity visual platforms: two of them used the same stock image!

The point is that companies and corporations should perhaps stop for a while the production of products and mediocre PowerPoint presentations, and instead start producing ideologies and true vision. True leading enterprises did it in the past. Can manifestos help to reverse the current state of affairs? I believe so. This is where the study of Futurist copywriting techniques for the production of manifestos comes in handy. In the chapter "Violence and precision: the manifesto as an art form" in her *The Futurist Moment* (1986), Perloff described some of the critical features that made the 1909 *Futurist Manifesto* by Marinetti such an innovation in the field of modern art. It has a number of technical elements that should be interesting for brand marketers and copywriters. For example, the internal features of the *Futurist Manifesto* are a unique semantic combination of narrative elements – this is where we find the contrasting clash of different styles and a literary technique that goes way beyond pure advertising slogans. The point is that Marinetti's *Futurist Manifesto* was not written as the simple enunciation of a platform of concepts: it is narratively entertaining, and it complements the ideas and the ideology with storytelling and gusto. In his correspondence with Futurist artist Gino Severini, Marinetti clarified the two key points of internal strength of a manifesto format:

- *A mix of iconic and bold statements where semantic effectiveness meets formal synthesis*: The point here is that there should be density and there should be conciseness in order to ensure that the key messages reach the mind – and especially the heart – of the audience in a memorable fashion.

- *Precise action against the "passéists"*: This last point refers to Marinetti's talent to provoke and to inspire debate and polemical buzz. It is important to go beyond political correctness and ultimately mediocrity by taking a considered position and stating a very strong viewpoint.

Describing the *Futurist Manifesto* of 1909, Perloff interestingly identified in the symbolist narrative that opens and ends the document the source of a unique power for an art manifesto: the power inducted by the "suspension of disbelief." What is generally reviewed as a point of conceptual weakness – the tension between the ultramodern message and the decadent setting where the action takes place – is appraised by Perloff as one of Marinetti's master strokes as a publicist – the talent to create a world of narrative where everything can be believed. This included the existence of the Futurist movement, a movement that up until the publication by *Le*

Figaro of this manifesto was actually limited to one member, Marinetti himself!

Futurism almost made the ability of its members to write manifestos into a specific conceptual art form, because by means of one or more manifestos, what was purely an idea became an artistic ideal, and in time and history, an organic vision capable of lasting 100 years. It is a structural belief beyond this chapter that these techniques fall into what we call nowadays "brand management," as then intuitively applied to modern art movements. Hence, the universal validity of the technical writing methods to create this powerful format. Of course, our challenges do not apply just to copywriting; they apply to our talent to tap into our culture, because what mattered to Marinetti's success was that the ideas of Futurism were in line with the culture of their age.

Marinetti achieved his "Futurist moment" thanks to his sheer intuition: how will contemporary brand marketers find their way to tap into the power of the cultural context of our own times?

Finding 14: Create your brand by anticipating cultural trends

Tapping into culture for brand marketing success

Think how so much marketing is simply fake, artificial, and superfluous. The fine arts, the cultural sector, the lively dynamics of urban design can offer the sources and the means to change this. We saw earlier how Mona Lisa manages to plug the power of fine arts in the context of vision and business process. It is now time to look at three examples of how fine arts can underpin superior brands, marketing, and design. Our first example will keep us very close to the earlier historical "findings" in this chapter: it is a very niche-oriented case history, not for the masses, yet inspiring to some.

Starting from modern art, a new format for integrating museum-level fine arts into the everyday is being designed and deployed in Milan, the city that saw the birth of Futurism. This project is (not by chance) being pursued by ISISUF (Istituto Internazionale di Studi sul Futurismo), the cultural association founded in 1950 by the last poets and painters of Futurism, to sanction the end of Futurism as an active force in the fine arts and nurture it as a historical one. ISISUF is an active cultural entity, conserving and promoting the archive of important modern artists, such as Carlo Belloli or Brazilian sculptress Mary Vieira, whose retrospective was organized in Sao Paulo and Rio de Janeiro in 2005, in the prestigious CCBB (Centro Cultural Banco do Brasil). In 2009, to celebrate the centenary of the birth of Futurism and to embody its original spirit, ISISUF launched the FuturDome concept to potential investors and partners. It is positioned as the first museum you can actually inhabit. Displaying key pieces from the unique ISISUF collection of modern art and design, FuturDome will transform a restored Liberty building (in a style from the early 1900s) at the heart of Milan city center into a permanent display of masterpieces, both known and rare, by Marinetti, Depero, Belloli, and many others. This formula will bring a unique edge to the apartments designed by cutting-edge, contemporary architects and emerging talents. The art of Futurism and its natural radiance will once again become part of the everyday life of the residents. To further widen the uniqueness of this interior living experience, those who rent and live in FuturDome will have to open their flats and apartments to visitors and art lovers for specific events organized by ISISUF.

In keeping with the spirit of Futurism, FuturDome brings back the art experience to the everyday, and expands its reach from the notion of museum visits or private collections to a new level of experience. Of course, once again, we are here examining a very specifically high-end targeted proposition: perhaps closer in its scope to Fondazione Prada or its equivalent in hospitality than to mass marketing. This is however a powerful branding opportunity for companies and corporations willing to associate themselves with fine arts and historical design selections, both by means of events to be held in the adjacent showrooms and through the product placement opportunities within each apartment, from utilities to accessories.

From the context of upstream real estate, we move on to the vibrant jolt of the urban streets, and from Milanese chic to the favelas (shanty towns) of Rio de Janeiro and Sao Paolo. This is where the contemporary new city rises, to refer to the title of a famous Futurist painting by Umberto Boccioni in 1909 (*The city rises*). Boccioni represented the energy of 1910s Milan by portraying the muscles and the sculptural structure of a horse in a building site in progress. Today, it is Rio de Janeiro and Sao Paolo that are home to some of the most innovative and astonishing manifestations of street art in the world. Such street art is authentically Brazilian, and representative of the way young people live in the favela. This is the true and authentic culture of Brazil today, seen from the talent and the passion of those who make it happen, against all the odds, mostly as a way to escape the inevitability of crime or poverty.

Until a few years ago, not many Brazilian street artists could offer their talent on the international market. This meant that their aesthetics and their vision of the world were confined to the cities where they initiated street art projects. Caramundo, the Dutch/Brazilian art foundation that connects underprivileged Latin American talent with advanced economy creative industry customers, changed this. The mission of Caramundo is to tap into the hip-hop, street fashion, and graffiti scenes of Brazil, and help bring all this energy and beauty into a constructive, sustainable, and repeatable business cycle. This translates into western companies reaching into the unique talent of trendsetting metropolitan South America. We return to Caramundo in Chapter 8.

Tapping straight into the world of emerging or cutting-edge fine art is an approach to the creative side of marketing communication that is increasingly pursued by more independent creative agencies. In a sense, the finest small firms do this. A perfect example is 515 in Turin, the viral marketing collective behind important advertising campaigns for the likes of FIAT Group Automobiles, which includes the world-class FIAT 500. Born as a team of designers and architects with a strong presence in the fine arts sector, 515 managed to express creative leadership by producing advertising that offers an opportunity for large corporations and cutting-edge brands to tap into the talents of visual artists and writers. For automotive fairs, 515 hired underground artists to contribute to the stage experience of Lancia, the upstream FIAT brand. It hires nationally renowned writers such as Giuseppe Culicchia to brainstorm and then write copy for sophisticated viral marketing campaigns which have experienced great success on low budgets. This principle applies both to large multinational brand campaigns, and to the campaigns of cutting-edge brands such as Ter et Bantine. An avant-garde Italian fashion icon, Ter et Bantine has a history of conceptual communication dating back to the 1990s, when Franca Soncini, one of Milan's leading fashion PR companies, established a connection between its brand communication and the visual world of cutting-edge contemporary culture. Nearly 15 years on, 515 continued in the same line by choosing Turin-based photographer Monica Carrocci to develop the brand's visual message.

Using fine arts content for the purpose of corporate communication is not new. What is new and of the utmost importance is the authenticity of the operation. There is a fine balance between the simple, detrimental exploitation of an artist's name or of an entire fine arts legacy, and the actual appreciative and appreciated meeting of an artist's talent and a brand. Examples of the latter are the work of Caramundo in Brazil and

515 with Ter et Bantine. We mentioned an example of the former in Chapter 2, the Chanel roadshow of art based on designer handbags. A perfect example of how art integrity and provocative marketing can successfully coexist comes from the museum world as we discover in the next "finding."

Finding 15: Create memorable encounters of tradition and provocation

Finding a fine balance between classic and contemporary

Futurism was all about the clash of the avant garde and the classic, by means of outstanding propaganda. How would such an approach look nowadays, in both curatorial and in marketing terms? Let's start this "finding" with a rather bold statement. Although you might not know him by name or would not recognize his face in a photo, Jan Willem Sieburgh is an icon. At least, this is the conclusion if we look at his portrait on the cover page of the Amsterdam's *Tijdschrift voor Marketing* as "Marketing Man of the Year 2008." The pleasant middle-aged Dutch businessman is a professional with a brilliant track record in the marketing industry, having formerly been director and partner of TBWA, one of the best advertising agencies of the 1990s. However, it is not his work in advertising that got Sieburgh voted as the best marketing man of 2008 in the Netherlands, it is his achievements in one of the most traditional institutions in the country, and one of the most famous museums of Old Masters in the world, the Rijksmuseum in Amsterdam.

Rooted in the Dutch National Gallery created in 1800 by the French on their own museum model, the Rijksmuseum belongs to the DNA of the Dutch spirit, in a display of symbolic power that stretches from the red, white and blue flag to the great water management works that keep the country's land dry. Hosted in a building designed and erected by Pierre Cuypers in 1885, the Rijksmuseum is at the heart of Museumplein in Amsterdam, and together with the Van Gogh Museum and the Stedelijk Museum, it is one of the great attractions of the city for worldwide tourism. In these difficult times of economic crisis and identity challenges, the Rijksmuseum is a solid reference point. Although most of the building is undergoing heavy renovation that is projected to last until 2013, the limited venue that is currently open to the public, showing work by Rembrandt, Vermeer and other Old Masters, and applied art including doll houses, is visited by crowds of foreigners in admiration, local students in excited confusion, and art lovers in ecstasy.

From the severe and serious pages of art history to the gloss of the Netherlands' most popular marketing magazine, Sieburgh (who has since 2002 been the managing director of this important Dutch institution) earned his award from the Dutch marketing and advertising communities with a number of brilliant operations. These ranged from the viral campaigns about "death icons" in the Rijksmuseum, to be "enjoyed" on YouTube, to an unique retail operation and annexed museum at Schiphol Airport, an experiment of transfusing art into the everyday transit zones of intercontinental passengers, with no equal in the world. Sieburgh managed to further innovate the brand marketing of the Rijksmuseum by turning the institution into a "media company-modeled" operation. He initiated, for example, a national glossy magazine, *Oog voor de kunst*, that won the hearts of the curators first, then won over the customers of newsstands and news shops. Here, the unique wealth of content available to the Rijksmuseum becomes an editorial asset, helping the Rijksmuseum to reach the masses.

In another part of his strategy, accentuating the popular orientation of his marketing vision, Sieburgh worked in collaboration with Dutch mass retailer HEMA, creating a line of co-branded popular items: cups, ceramics, and other similar mundane, everyday objects. At Hema, humble objects such as cups or mats become the brand focus for the wealth of the Rijksmuseum's visual culture, and the millions of Dutch people who visit shops every week.

The greatest achievement of Sieburgh however was not a pure marketing operation, it was a curatorial project that lengthened queues at the Rijksmuseum and encouraged sophisticated art critics and ordinary people to talk more about the museum. Sieburgh's idea was, in retrospect, simple: get one of the most representative icons of today's art, identify a specific object that would stand for the synthesis of contemporary sensibility, and from there, build a conceptual clash of the great Dutch Masters with the cutting edge of the 2000s. It's brand marketing by shocking, you might say. In order to achieve this, Sieburgh chose artist Damien Hirst and his *For the love of God*, a skull modeled in platinum and encrusted with diamonds. This piece is estimated to be the most expensive work of art ever. The operation entailed putting this unique contemporary artwork in the sacred rooms of the Rijksmuseum, with a show opening on the Day of the Dead, an entire issue of *Oog* magazine focused on death, and great media coverage. Sieburgh went one step further: just before the immersive, dark space displaying Hirst's skull, he showed 16 masterpieces from the Rijksmuseum collection as selected by the British *enfant terrible* of contemporary art. More traditionally oriented curators were not pleased by the lack of historical consistency in Hirst's selection, and some were irritated by the apparent lack of depth in his commentary notes. And of course, as usual with Hirst, the marketing machine of gadgets and memorabilia was ready in a dedicated part of the museum, with premium prices for items featuring the diamond-plated skull, from T-shirts to pencils.

The impact was world class in quantity and quality, in press coverage, audience enthusiasm and crossover buzz between classic art aficionados and contemporary art fans. Sieburgh identified an icon of our times, a controversial artist with a high media value and a highly provocative legacy, and saw the opportunity to host him in the austerity and dignity of one of the world most serious art institutions. It would not have been any use to Hirst to display his skull in a contemporary art museum. That might even have been detrimental to the charisma of the "most expensive art object ever." Sieburgh offered Hirst the unique opportunity to engage in a dialog with the Old Masters, and to make a further step in his personal exploration of the themes of death – themes that Sieburgh himself developed in his marketing campaign content.

The cultural clash here happened on several levels:

- the contemporary art edge of Hirst versus the iconic and symbolic dimension of the Rijksmuseum in the Netherlands

- the postmodern curatorial and marketing machine deployed by Hirst versus the philological authority of the Rijksmuseum

- the iconic presence of the skull versus the presence of the Old Masters, and the museum experience before and after the Hirst rooms.

The lesson is that innovative and daring combinations of the old and the new, the sacred and the profane, the shallow and the deep, do not only generate controversy. Such combinations generate popular interest because they hit the imagination of a wider audience; because they reach out, and are memorable, to the majority. This is a challenge and an excellent opportunity for all those brand marketers who prefer the average safety of mediocrity, to dare to explore new opportunities.

Wrapping up: from manifestos to sharper marketing

To capitalize on and capture the "findings" presented in this chapter, let me mention one more communication design agency, this time a rather specialized one in the cultural sector. Thonik of Amsterdam is a "boutique business" hosted in a rather controversial "shoe box" Amsterdam building by MVRDV architects and interior-designed by Jurgen Bey. Thonik is the creator of award-winning campaigns for leading museums such as the Boijmans van Beuningen (Rotterdam). This is a leading agency in terms of understanding the unique way the cultural sector works, exploring innovative brand communication mixes that make the difference. How does Thonik work? Its process can be explained as follows:

- First, extensive analysis is performed to understand the world in which the brand exists, and to extract and/or provide context to its values.

- When the analysis outcome is acquired and digested, a basic design framework is established, in order to provide a solid backbone to the project.

- At this point, total freedom of exploration comes in. This does not simply mean establishing a media-neutral communication strategy: it means a "blue sky" creative journey, where no fixed tools or evaluation processes inhibit the unleashed power of generating ideas.

One of the most praised and awarded campaigns by Thonik using this process was created in the national political arena. The Socialist Party (SP) is a former Maoist political group of the Dutch left wing. Its platform for the recent political elections was simply "Better Netherlands start here," with the slogan: "Now SP," and a Thonik-designed logo representing a star as part of a bright red tomato. As well as this logo, Thonik created a number of events designed to have a big impact. A Dutch design classic is the SP soup bowl, containing a specially made tomato soup with the signature of a celebrity Dutch chef, which was distributed to passers-by in Dutch cities. Thonik also created the concept and design for an entire stage performance accompanying the closing speech of the party leader, with a gospel concert complete with a chorus dressed up in long white aprons, displaying, of course, not the signs and symbols of any religion, but the red SP tomato. This campaign by a party inspired by Marx's manifesto resulted not only in SP's relative success in the elections, but – quite in contradiction to an antagonist political vision – in universal praise and the granting of an award by the Dutch marketing community. Thonik showed how the lessons learnt by excelling in the brand marketing of fine arts translate into success for branding anything, from products to politics.

What if we were to capture the essence of how corporate managers and brand marketers might want to rework their processes, in order to rethink their work? The "five findings" in this chapter can be summarized as follows:

- Dare to dare, and aim to reach the world even if your means are modest and your assets seem irrelevant: with the right propaganda strategy, you will stand a chance.

- Aim to express a vision of the world and beyond, a vision of the universe. Do not limit yourself to your category or your niche.

- Do not produce just products or brand PowerPoints: people want more from brands. Produce an ideology, create your own vision of the world, and translate it into a sound manifesto to reach out to all audiences.

- Stop working with standard marketing communication processes: look at the culture that makes the world we live in, immerse yourself in it, and use those agencies and agents that can connect you to it.

- Dare to dare when it comes to creating a short cut between the contemporary and the classic, the past and the future.

In the spirit of this chapter, we can notice that although Thonik relies on a process, it is a process designed to open the creative phase to what might look like complete anarchy to ISO design managers or marketing planners. This kind of anarchy is the way in which this specific agency, and other creative boutiques, manage to make a difference, and create unique value. Because agencies like Thonik or 515, or foundations like Caramundo or ISISUF in their curating/consulting capability, are grounded in the body of cultural streams and trends, their processes will inevitably be more "relaxed" than those deployed by McKinsey and other consulting and/or advertising giants. It is a matter of leadership and authenticity, and intimacy with artists and with the creative edge. It is also a matter of a relationship of trust and intimacy with customers who often are mavericks themselves – borderline cases of corporate management or cultural sector management who bend the rules to the goals, and consider such goals more relevant than the means.

This is the main message of this chapter and this section: beyond process and conventions, marketers need to go back to the soul and the roots of what brands stand for. More than ever, marketers need to reconnect to our culture and the cultures where they want their brands to thrive and prosper. We will need to liberate the best creative minds in marketing, advertising, design, innovation, and all other agencies and departments, in order to move beyond the rigid frameworks and the protocols of the twentieth century, adapted and deployed for the new digital age by technocratic administrators. The economic downturn will not go away with sterile textbook formulas, just as the twentieth century did not retreat when the traditions of the past were challenged by artists such as the protagonists of Futurism. The manifestos of the future should be written from scratch. It is our choice whether to be the ones to do this, or passively wait for history to make us irrelevant.

What is Futurism?

2009 was the centenary of Futurism, the Italian modern art movement that spanned nearly 40 years of active history, and then influenced the rest of the twentieth century more indirectly. Its founder, F. T. Marinetti, and some of his most talented followers, established a powerful grammar of multidisciplinary fine arts and applied arts. Coming from Italy, which at that time was a recently united, culturally provincial country, they established a path from nothing to fame to decadence, to universal recognition and today's contemporary celebration. The program of Futurism entailed a violent rebellion against the art and the lifestyles of the past, with the energetic pursuit of aggression, speed, and dynamism as the leading traits of a new aesthetics for a young art.

The modern city, industrial landscapes, and the beauty of the machine and of the automotive experience were among the reference points for Futurism. Praised by communist thinkers like Antonio Gramsci and adopted in its aesthetics by Russian revolutionary artists, the movement was positioned for some years at the crossroads of political left and right wings, to then finally lean and fall into the reactionary camp of fascism. The international nature of Futurism ensured that Marinetti, as a true internationalist, would never make it into *art de régime* for Mussolini, and kept Marinetti strongly adverse to any racial discrimination, especially the shameful 1938 Italian racial laws. Prior to the 2009 celebrations (started in 2008 at the Centre Pompidou in Paris), Futurism was marked by the stigma of its ideological connections with fascism, although it had always been recorded as a primary movement of Italian modern art, the only really international one before the 1960s Arte Povera. As captured by the archive of ISISUF in Milan, the innovative power of Futurism at its best is universally acknowledged, today more than ever. Such power to think "blue sky" and to reach the masses by means of propaganda techniques is the thread that links the inception and launching of the movement in 1909, through the 1930s aesthetic experiments in "lower" culture, like advertising and design, aiming to (re)create not just painting or fine arts but the totality of a contemporary universe.

PART III

Relationships and Museums

In Part III we move our focus to the social, institutional, and commercial sides of the fine arts. Starting from the commercial side, it is a given fact that, as much as we might discuss fine arts from various philosophical viewpoints, its financial viability and economic base of existence ultimately determine to a large extent its ideologies at any time. The art market is where artists and movements find their own growth areas and their best milieu to prosper, both economically and intellectually. At the time of writing, the global art market is exposed to the fluctuations and disruptions of the economic downturn – just as any other premium value-driven sector is. There are however exceptions. As an example, the TEFAF 2009 art fair of Maastricht, one of the most relevant art events in the world, confirmed its leadership within its category with a stunning preview day. The success of TEFAF in times of financial depression is not a surprise: it could actually be a counterintuitive confirmation of the difficulty in the economic situation. After all, TEFAF is in the premier league of art fairs: the equivalent of Cartier or Hermès in the fine arts fair category. It is therefore no surprise that in the current situation, those wealthy people still willing to invest in fine arts and luxury selections concentrate their spending power on the most prestigious, the most solid, the most reliably established brands in each category.

This section investigates the milieu where art vision translates into sales of artworks, and the professionals and institutions that support such markets, either indirectly or in immediate terms. Museums are reviewed as key cornerstones of the overall fine arts system. This analysis reports a number of traits regarding their institutional function, architectural presence, and business models. We investigate the meanings of museums from sociological, commercial, and design viewpoints. The questions addressed by this part are:

- What can business players learn from the world of fine arts communities and collections?

- How can contemporary museums inspire strategy and marketing visions?

In Chapter 5 ("The art of relationships: from communities to collections") we introduce a number of fascinating stakeholders of the fine arts system – dealers, gallerists, and especially collectors. They are guides in discovering the peculiarities, the rituals, and the rites of the art market. The relationship between the integrity of communities and the success of the art markets is a very delicate territory to explore. Here, brand marketers and business leaders could be greatly inspired by the complexity and the richness of interactions among artists, publishers, gallerists, collectors, and museum managers. We meet art critics such as Germano Celant (director of the Fondazione Prada in Milan), gallerists such as Antonio Tucci Russo (one of the key enablers of the legendary Arte Povera movement), and traders such as Seth Siegelaub (who in the 1960s managed to promote and market an art movement delivering ideas and concepts, and therefore without real "objects" or artwork to sell). Our case study looks at the psychology of a fictional collector, created by the very much missed author Bruce Chatwin. We conclude Chapter 5 by observing the conversion of collectors and basic art lovers into patrons of museums. A deeper examination of the dynamics behind this crucial institution is in Chapter 6.

Be it the role that museums play in terms of urban planning or be it their cultural mission, there is no doubt that the idea of "museum" has been enriched, in the last few decades, with multiple meanings and multi-layered depth. As presented in Chapter 6 ("Between 'ideology' and space: the museum experience") museums are entities worth further analysis, at both the cultural policy and the business modeling level. Museums are important not only as city marketing icons, architectural masterpieces, and ideological think tanks, but also in terms of pure enterprise and commercial success. We find evidence, in the policies and in the practices of selected leading museums, that commerce is definitely not bad for culture, and the same is true the other way around, if the mix is properly managed. This is confirmed by our case study, where a socially engaged museum is discussed, with its policies matching a high degree of commercial drive. In this chapter we meet architects Frank O. Gehry and Yoshio Taniguchi, and visit prestigious museums such as the Louvre (Paris), the Museum of Modern Art (New York) and provincial gems such as the Centraal Museum (Utrecht) and the Prefectural Shinano Art Museum (Nagano, Japan).

In this chapter, business readers and brand marketers can find synthesis and inspiration for reflection in the field of vision and identity design, corporate estate, and branded retail, and business innovation and cultural consulting. This Part has ten "findings" in Chapters 5 and 6:

- Finding 16: Engagement at its peak: how art communities were born.

- Finding 17: Engagement as the engine of commercial growth.

- Finding 18: Engagement and enterprise: corporate collecting.

- Finding 19: Discovering the complexity of the most engaged customer in the world.

- Finding 20: From personal engagement to advocating culture.

- Finding 21: Learn who you are by searching your soul.

- Finding 22: Create your brand identity to express your personality flexibly.

- Finding 23: How buildings express what you are.

- Finding 24: Express your soul by connecting to your urban context.

- Finding 25: Express your personality by connecting culture and commerce.

In Part III, we explore how art communities are formed, how museums are managed, and how commerce can connect with culture in the most effective fashion, within the boundaries of integrity. These dimensions of integrity, authenticity, and genuine respect for artists and curators are essential to search and maintain the soul of the fine arts while commercializing them. Part III demonstrates and reinforces this principle, while presenting the forms of profound engagement that are characteristic of fine arts collectors. These are by far among the most intense, passionate, and driven customers in the world: how can we miss the opportunity to learn more about them? We address the usual key focus of our "findings" – primarily corporate business leaders and brand marketers – while at the same time looking at considerations that apply to the work of museum directors and cultural sector managers.

CHAPTER 5

The art of relationships: from communities to collections

Since the rise of digital marketing and in particular since the explosion of Web 2.0, "community building" has been one of the buzzwords in every possible business debate. The social change in behavior resulting from co-creative participation of people in communication channels resulted in a parallel ambition to commercially exploit these new cultural trends. Unfortunately for most marketers, a community is not just a list of email addresses: it takes much more to build true human interaction, emotional bonding, and value over time. The connecting thread that runs through this chapter is "How can we perform successful commerce by nurturing genuine communities and respecting their integrity?" To answer this challenging question, we look at the uniquely intense passion of collectors, also known as "possibly the most eager and committed customers on the planet." We then move to looking at corporations in their own "art customer capability," in order to gain an understanding and insight into how collections might work for the best benefit of brands. Our last "finding" introduces us to museums (the topic of Chapter 6) from the viewpoint of their connections to collectors who have become generous art patrons. Together, art communities and art collectors offer us insights into how to build a bond with customers by genuinely being part of their social circles.

The business question behind this chapter

What can business players learn from the world of fine arts communities and collections?

An introduction to the findings in this chapter

Corporate management and enterprise leaders normally have the explicit ambition to create long-term relationships with value-driven customers. Collecting is a commercial territory where such a game has been played for centuries, at unprecedented heights of sophistication. Here, the creation of relationships and the consistent management of them over a long period are essential. Collectors will go a long way to acquire and possess the works they deem necessary to complete their collection. As demonstrated in our case study, collectors can go to extremes to create a world from their collection where they find themselves more at ease and at home than in the real life. To its owner, a collection is a living organism. Beyond any consideration of economic nature, in terms of investment or value, the profound emotional connection between collection and collector is one of the highest peaks of bonding between humans and artifacts. If we imagine an ideal zenith at the heart of our golden crossroads, this is at the point where communities of artists meet collectors.

In the last 20 years, the value of community building has been analyzed and advocated by marketers all over the world. Nowadays, being part of a community mostly means to be registered in some sort of online archive, sharing content, and being reachable. The question however is: To what extent should we speak about just connectivity, and to what extent should we instead speak of a real community? The answer from the world of fine arts is simple: true art communities are built by a number of stakeholders with different roles and different kinds of contribution, but with one strong element always in common – a high degree of intensity of participation. Art community constituents organize themselves around a shared idea (or, even better, ideal) with the strongest drive. It is this intensity of contribution and participation that business leaders and brand managers should aspire to at all times in their customer relationship management.

Understanding how art communities and collectors think casts a whole new light on the emotional relationship between your potential customers and your brand in terms of symbols and meaning. From erotica to comics, people collect anything in the world: one of the key questions you will want to ask yourself after reading this chapter is: What is my brand missing that it requires in order to become collectible? Of course, to high-end brand marketers this should be a burning question more now than ever. As any luxury goods marketer knows from the trade, one of the most important factors in premium value markets is that whatever the transaction is, the goods or services purchased translate into a higher degree of social prestige. Such social prestige (surely one of the factors of pride of every collector) is however not based just on possession of the goods and showing off. On the contrary: within art communities, true prestige comes with being part of the right peer group and sharing the same level of sophistication. Within contemporary art, there is an even higher degree of participation that marks the difference between the nouveaux riches and the true, committed, prestigious collector. It is the community level of those collectors who join an art movement at the very start, and invest not only their financial means but also their reputation and – we could say – their belief in nurturing it. Those particular collectors become part of the community behind and within the movement, and are blessed by unique insights in an in-depth debate with the artists. Intellectual intimacy and existential empathy are essential to enable such debate. To some extent, these mechanisms could be replicated in the context of high-end consumption as well, if their past and present meanings are properly understood.

Before we further analyze the ways of working of contemporary collectors, we should ask ourselves: What are the roots of art collections, historically? According to scholar Krzysztof Pomian, collections as we define them today were born in the ninth century, in Venice, around the sacred remains of St Mark. The rulers of Venice created a collection itemized according to a political and religious agenda of power. Based on this original model, centuries afterwards, in the Italian context, the notion of private collection as we know it was born. Only in relatively recent times, corporate collections followed and established themselves at the highest level of market relevance. We further analyze these historical evolutions in this chapter.

In this chapter we first focus on dynamics in the arts sector described above: we start from communities of artists, we move to the markets of arts and the key role played there by collectors, and conclude with the mechanisms of art patronage. The main points of this chapter are:

- *The integrity of art communities, and why integrity is essential*: This is a crucial lesson for any marketer willing to engage with and become part of a real social circle: here, genuine participation is vital.

- *The conversion of fine arts movements into commercial success*: This is where fine arts show how true marketing talent translates into the ability to sell even the most difficult propositions – for example, conceptual art – if marketers operate strictly with respect for their integrity.

- *Pros and cons of cultural investments*: This is where corporate management and business leaders can gain insights in how to increase their social capital in terms of acquiring a closer relationship with higher culture, by initiating and managing corporate collections.

- *The life journey of a collector*: We track the steps of a collector from early involvement with the art market to maturity, in order to understand the collector's world better.

- *Collectors, museums, and societies in debate*: We conclude with the conversion of informal business networks and communities into a sponsorship vehicle for museums, and the universal validity derived from this particular context.

Business readers are invited to observe how the mechanisms of art markets and communities work. This should inspire them to consider innovative approaches to their own trade: sometimes at niche level, sometimes at the level of general principles. This chapter offers inspiration for the reformulation of the framework where marketing takes place: the purest relationship of attraction among people and their objects of passion from the viewpoint of both owning sophisticated artifacts and being part of social circles and communities.

Finding 16: Engagement at its peak: how art communities were born

Understanding the reality of genuine community building

In Chapter 4 we looked at the history and development of Futurism, the great pre-Second World War Italian art movement. We move to the fall of 1967, and encounter the second great period of recent art in Italy, labeled by Germano Celant (future director of the ground-breaking Fondazione Prada) as "Arte Povera." Arte Povera was a human-focused, non-objectual current of art that interpreted the spirit of the late 1960s, introducing "poor materials" such as cartons, mirrors, rags and similar into the full spectrum of contemporary aesthetics. An offspring of the city of Turin, Arte Povera established and represented a whole new sensibility, one drawing from students' movements and from the rising wave of left-wing protest that would translate into a 1970s ethos. Although philosophically, politically, and aesthetically the antithesis of Futurism,

Arte Povera expressed an equivalent level of artistic international excellence, to the extent that these two movements represent the pinnacles of Italian art of the last 120 years.

Celant launched his conceptual platform defining Arte Povera in a groundbreaking 1967 article about young artists from Italy. Titled "Notes for a guerrilla," it was published in the then cutting-edge *Flash Art* magazine. Where nowadays you find the vibrant congregation of trendsetters and early adopters around dynamic blogs such as www.curating.info, www.wooster collective.com, and www.trendbeheer.com, in 1967 *Flash Art* stood as one of the best antennas to discover the cutting edge of curatorial and ideological developments in contemporary art. *Flash Art* was, and has been for at least 25 years, the gatekeeper of an Italian community of artists, dealers, scholars, collectors, and curators who defined the trends in their respective fields. Celant, not unlike Marinetti in 1909, chose his publication well in order to launch his message to the world.

As an almost natural second step, Celant established the territory of Arte Povera by curating an important group show, under the title of "Conceptual Art Arte Povera Land Art" at the Galleria di Arte Moderna (Turin), in 1970. These trends were not easy to present: for example, artworks were being absorbed into landscape, in the form of interventions in the desert (Heizer, De Maria) or in the trees (Penone). Art objects were realized in materials of great simplicity, and almost dismissive, humble everyday modesty. Ideology, much more than anything else, prevailed in determining the value of an artwork, with just certificates or documentary records (like photographs) recording the actions of artists. From Arte Povera we can take two important moments in the public launch of an art movement:

- The publication of the "foundation bible," or the manifesto that describes its perimeter, or the program built around ideas, as described in Chapter 4 on Futurism.

- The actual expression of such a program in a masterpiece selection compiled by the leading expert. This should be at best a collective show of potentially historical impact curated on behalf of a trendsetting museum or art institution.

How do critics discover new communities and new generations of artists as Celant did in 1967? In general, we could say that today the spontaneous, viral meeting of young artists still happens through schools, informal contacts (for example, nightlife), and sometimes research-oriented galleries. Critics review their work at early stage, clustering it into themes, eventually leading to a "foundation" basis for a movement. Between the foundation and the first important collective show, important intermediate steps have to take place. For example, an important strength of Arte Povera was its social dimension of informal networks. This was a mission-critical factor to gain wider market acceptance and commercial success. If we describe this process generically (for now keeping a specific focus on interactions involving just artists, critics, and collectors) it could look like the following:

- After the first shows in underground settings, collectors emerge willing to risk and to invest in a new aesthetic language: these mostly belong to the same age group and therefore have the same taste and mindset as the young artists.

- The creation of a virtuous circle of interest and attraction, where cutting-edge publications (such as *Flash Art* in the 1960s or blogs today) acknowledge the works by young artists, reassuring early collectors in terms of the value of their investments, beyond any personal or emotional affinity, in a sphere of economic rationale.

- Then there is the time of validation, when a major institution (such as a museum, public gallery, or a biennale) takes on board the young movement, and the movement emerges from the semi-underground to become established, accepted, and profiled as one of the signs of our times.

After this, the community of collectors, critics, and artists is consolidated, and constitutes a historical moment of the art experience: new young movements emerge, determining the formation of new communities. New collectors within such new communities endorse new art. The role of these more open, most experimental early collectors is crucial to the economic viability of new art and its communities. These are the explorers, who buy with their heart and with their desire to be a part of what is going on. From a marketing viewpoint, they can be described as early adopters, but what is important here is the fact that (together with the artists) they form both sides of the art community coin. In the case of Arte Povera, but also in general terms, such continuity between artistic production and cultural understanding is unique in its generational dynamic, and strong in its aligned sensibility. It is a match that goes beyond the purely functional side of investment or adoption: It goes straight into the symbolic value of defining who we are as individuals in a given time in history.

One critical enabling factor is crucial to connect communities and collectors at this stage: It is vital to have agents in the system that are respected and, especially, trusted by all parties involved. This requires authority based on integrity and on the ability to combine intellectual sophistication with commercial pragmatism. The question then arises: Who plays the crucial role of these gatekeepers?

Now is the time for the entry of the third element in the business model of Arte Povera and of any young movement emerging in fine arts and design. After the art critic/curator, and the collectors, we have the gallerist. This is where we find the so-called "primary gallery" (the gallery that buys work directly from artists). A primary circuit gallery is a crucial part of the entire community system, as it acts as talent scout and mentor to the artist, while commercially channeling the proposition to the young collectors. For example, in the case of Arte Povera, the then Turin-based Antonio Tucci Russo was one of the gallerists who played such a role. The function of such dealer is not that of pure hard marketer. The sales process takes place (as business sustainability is necessary for the enterprise to remain viable), but the soul of the business is not in the financial accounting, it is in the cultural challenge to discover the new, to establish a bridge of discussion between artists and collectors, and to make the culture of their own times happen.

To many in the art world, this might read as an old-fashioned view. After all, until this current financial crisis, we lived in the age of star galleries and art management practices that resembled luxury goods commerce. There is a certain degree of disdain among the veterans of the art system for this process of extreme commercialization. There is also a clear view that (in the wake of the current global crisis) such practices will be strongly downsized and a new focus on quality of work will prevail again.

We can draw one preliminary conclusion of value: Marketers should accept the challenge of building a real culturally contextual bond with people, not just be satisfied with doing what is described in most marketing manuals, because in the longer term, such a bond will translate into superior relationships. In the case of Tucci Russo, the strength of his customer relationship management (CRM) approach has been such that, since the early 1990s, he could afford to move his own business from the center of one of Italy's art capitals, Turin, to the remote town of Torre Pellice in the Alps, close to France. Here, his coherent and consistent program continues to attract collectors, art lovers, and long-term friends. Imagine moving your retail enterprise from strategic locations in the high street or downtown area of cities to the mountains, and being virtually unaffected in your business operation: could you afford to make such a choice? If you are an average company brand manager or marketing VP, the answer is "no." However, if you are a cultural collector for the benefit of a "real" community, things change, and for the better. The next question is: How can a dealer, this time in a marketer role, facilitate and even stimulate the commercial success of emerging, unproven, "difficult-to-digest" art trends? In the next "finding," we unveil some of the secrets of the trade of art marketing, connecting them to the reality of marketers and their work.

Finding 17: Engagement as the engine of commercial growth

From community to commerce, respecting the integrity of fine arts

One of the fascinating aspects of modern and contemporary art is its pushing against boundaries. Contemporary artists have been, to a certain extent, rebelling against the viewer and even the potential customer, be it an individual or an institutional collector. Within such a context, in some cases, the challenge to market art works seems an impossible one. On the other hand, great challenges attract unique talents, and give the rest of us an opportunity to learn. As we mentioned, Arte Povera used materials of great formal modesty, and expressed a violent climate of anticapitalist protest. This resulted in a strategy designed to generate art that could not be sold by regular means. At an equivalent level of commercial challenge stood the art movements of land art and conceptual art in the United States. Here, a 1968 collective show at Windham College by rising stars Carl Andre, Lawrence Weiner, and others displayed works in the open of a purely ephemeral, fully contextual nature. A sculpture by Andre, *Joint*, consisted of a line of bales of hay connecting two green areas. The sculpture on the land was designed to self-destruct in the face of changing weather conditions – rain, wind, sun – and of the fully unrestricted passage of people. In line with such poetic directions, conceptual art established a new, nonobjectual, anticonsumerist platform of aesthetics, one where the "idea" would be the actual true value in art terms, while the "object" as the subject of a possible commercial transaction was deemed not of any value. Of course, this new kind of art still needed to be marketed: selling artworks that are nothing more than an idea was a challenge, especially in terms of contractual ownership.

Celebrated in a book by MIT Press, *Conceptual Art and the Politics of*

Publicity, Seth Siegelaub is described as one of the most influential protagonists of the late 1960s/early 1970s American art scene. He operated behind the scenes, leaving the stage to artists and to artworks. As acknowledged in an old interview by Lawrence Weiner, reproduced in the same book, Siegelaub was the marketing mastermind behind an entire generation of avant-garde artists who were difficult to exploit commercially. The end of the 1960s was a time of hope, growth, and experimentation. In this dynamic context – possibly the opposite of what we are experiencing nowadays – the United States consolidated its leadership as *the* scene where contemporary art directions were defined. In particular, New York was the city to be. And NYC was the networking hub from where the protagonist of this "finding" operated. In this context, Siegelaub, as a young man of 23 started up his first business – the Seth Siegelaub Contemporary Art gallery. This enterprise was to be short-lived, as the pressures from costs and overheads plus the opening of more and more new galleries made it impossible for him to remain viable. Siegelaub resolved to move his business to a small apartment at Madison Avenue and 82nd Street. It might have looked like failure, but it was the start of cultural growth and commercial success.

A pragmatic man of his times, Siegelaub stopped being a gallerist, and started to become a ground-breaking marketer. He operated by nurturing relationships with key people from his list of business contacts; he consolidated and expanded his social networks by hanging around in the right nightclubs and bars, and hosted private events, with an invitation-only policy, showing art and launching new artists, in his own living room. He used the same methods as the community building of Arte Povera, as described earlier. Rapidly evolving as a master of publicity, Siegelaub maximized his reach and his equity as a valued stakeholder in the community chain between new, experimental art and potential new, young collectors. Not only did Siegelaub leave his office to be with customers as every marketer should, he shut down his enterprise to go back to the roots of what the art dealing business is all about: being part of a community as the link point between cultural offer and commercial demand. How did he play such role?

Siegelaub did not intervene in the creative flow of the likes of Andre or Weiner. This would have not been thinkable. What he did was to engineer the whole marketing side of the proposition, starting from the contractual structure that would enable a sales transaction to take place even if an idea was the object of the contract. First, after discussing with more than 500 art world experts of all kinds, Siegelaub developed, in collaboration with Robert Projansky, the "artist's reserved rights transfer and sale agreement." This contractual format was a milestone because it enabled the feasibility of commercial operations for an art approach where the value of objects was ideally none, and the value of the freely transmittable idea was everything. Second, as the organizer and *deus ex machina* of the show at Windham College, Siegelaub deployed a number of interesting public relations techniques, which – even now – do not appear outdated at all:

- The show was not held in NYC, then the center of buzz, but in a relatively remote location: this decentralization offered the unique possibility to differentiate it. Just as with Russo's gallery, the location was a strategic winning choice under specific circumstances.

- The show was opened with a forum, moderated by upcoming trendsetter Dan Graham and with a contribution by the key featured artists. Transforming the opening into a public conference added substance and future possibilities of exposure for the featured work and its creators. Based on direct professional experience, there is no doubt that the organization of forums and symposia can prove of great value to marketers. With relatively small investment, a great deal of positive reputation is generated by a genuine, authentic, intellectually honest debate and discussion with top experts. Media coverage and viral buzz will inevitably follow.

- The extreme nature of the specific site and the ephemeral nature of the works installed, as exemplified by Andre's *Joint*, made the show not only an incredibly powerful statement of integrity and consistency but also a kind of mythical event from a historical perspective, along the lines of "You had to be there to see it." This could only raise the bar in terms of future charisma.

The last point is important in terms of storytelling and narrative. This curatorial approach created a degree of mystique around a project, where the intensity of site-specific intervention met the rigor of theoretical debate, with a number of documentary photographs acting as testimony of this pivotal moment in contemporary art.

It is time for some preliminary conclusions. It is an option for marketers to act as contrarians: just like Siegelaub (or Tucci Russo) did, even to the point of relocating or delocating their business. And especially, it is time to change the rules of the marketing game: as these pioneers of new art grammar did, it is time for business leaders to not just pursue what the market wants. It is time to decide instead what you want to nurture as *the* proposition you truly believe in, and then pursue it with all of your passion, reinventing the way business is done. Although it might seem like a paradox, co-creation can help to create a different future, if there is leadership with integrity. After all, isn't creativity the secret weapon fine artists and marketers aim to reach, to seduce and to truly engage their customers?

Finding 18: Engagement and enterprise: corporate collecting

New debate between fine arts and corporations

In Finding 17, we looked at the marketing techniques behind the success of conceptual art. With this "finding," corporations as potential collectors are our primary focus. Our link point between these two seemingly distant topics is once again Seth Siegelaub's work: taking a step in terms of commercial challenge, Siegelaub managed to package the manifestations of conceptual art for acquisition for corporate collections. In order to do so, in the late 1960s and early 1970s he operated another company called Image: Art Programs for Industry, Inc., in collaboration with a business partner who was a talented businessman and an important collector. This was an important turning point in bringing conceptual art to the market as a potential investment for corporate collecting.

Why would a corporate organization, largely built on the profit-making principle, become an active player in society trends and cutting-edge culture? The answer is vital to this "finding." There has been a lot of interest in the last few decades about the social and societal role of corporations at different levels. We should start from the fact that the 1960s was a time of growth, when corporations expanded the production and distribution of goods to create mass markets in all advanced economies. Optimism was the driver because material prosperity was the goal apparently within reach. It is important to underline how this specific climate in those years provided the right platform for the vision expressed by Siegelaub. In the 1960s, corporations were however lacking in one aspect of their profile – cultural capital. As much as their wealth was growing in material terms, their reputation at intellectual and social levels needed injections of prestige. The prestige of being an actor in high culture and fine arts was the way for them to acquire social capital. Siegelaub understood this trend of corporate development at a very early stage, and perhaps at the time of most opportunity. He created a proposition to exploit this new demand, by servicing corporate sponsors willing to associate themselves with the best emerging art. The unique selling points (USPs) offered by Siegelaub's Image company to corporate collectors were:

- The opportunity to position the corporation as progressive, building up prestige in terms of reputation within diverse communities.

- The importance of reaching higher segments of society by connecting with the field of contemporary high culture.

- The chance to jump on the bandwagon at the optimum moment, when public interest was at its peak, while competitive advantage could still be built by early movers into this relatively new field of contemporary art sponsoring and collecting.

This short list of marketing and PR benefits remains largely valid for prospective corporate collectors even today. With the exception perhaps of the third point, corporate collecting has become, in the last decades, a widely diffused practice of great relevance for all involved stakeholders. This was already clear shortly after Siegelaub's start-up: just a few years after the launch of his company Image, Jeffrey Deitch, then a brilliant 20-something art advisor for wealthy customers of the US banking system (and nowadays a leading gallerist and curator), wrote a groundbreaking article for *Flash Art International*. The title of the article, "Art from the last creator of images to the last consumer," said it all. Its placing in a specific section of the magazine, "Art as investment," made clear to its audience that the focus was not on aesthetics or history of the fine arts, but on monetary value and financial transactions. In the article, Deitch described the PR value of corporate collecting with great clarity. He used the example of how major corporations benefited systematically – at reputation level – from cultural sponsorship. Deitch wrote that major PR firms already had in 1980 a dedicated department for fine-arts-related operations. Seen through Deitch's words of 1980, Siegelaub's 1970 vision became almost a prophecy.

Moving from history to today, the relationships between corporate players and fine arts evolved from just collecting to the advent of branded museums. Affiliations such as those of François Pinault, owner of the Gucci Group, with the Palazzo Grassi in Venice, and the creation of new private museum buildings as planned by Bernard Arnault of LVMH, could offer a natural transition from corporate collections being displayed in offices to the brand being at the center of an art experience. The main challenge here is to maintain the right balance of integrity in curatorial policies. To prevent detrimental PR, the relationship between corporate collectors, corporate sponsors, and fine arts curators needs to be inspired by a constantly renewed spirit of integrity. In line with this, looking toward the future, three additional directions of prospective change in the relationship between fine arts and corporations can be highlighted:

- Corporate collection curators will increasingly specialize, acting as custodians of the collection while becoming a more important part of the strategic staff advising senior management about cultural marketing and PR opportunities.

- Regarding the level of display in corporate headquarters and similar locations, new office architecture these days demands transparent environments and open space formats, with a great degree of flexibility and mobility. Classic fine arts collections might not be as well integrated into the office space than they were before the IT revolution. This will inevitably influence curatorial choices.

- From simple display within standard office environments to branded foundations and even museums, the purpose and ways of using the corporate art collection will continue to evolve. For example, imagine that fine arts could be displayed within branded flagships and retail areas in general as a standard practice in a few years from now.

Of course, there should be no mistake about the fact that corporate collectors are vital in making art markets. It must however be reiterated how fine arts have much more to offer corporations than just ownership of super luxury goods in the form of collection pieces. In the wake of the financial crisis, instead of just commercial relationships and limited transactions, it is in the best interests of corporate players to involve artists, designers, and architects at the very center of their business processes. We saw how Mona Lisa of Paris is a consulting firm that is already doing this with success. This is a cultural revolution that requires corporate leaders to be braver than ever. It is about being intellectually honest, having the courage to put creativity at the center, and aiming for real change. Corporate players may take the same attitude to benefiting from art that private collectors have: an irresistible passion that changes people's lives.

Finding 19: Discovering the complexity of the most engaged customer in the world

Discovering the complexity of the collector's passion

Collectors are complex people, both at the intellectual level and business-wise. This "finding" explores this complexity, and (in the light of the economic crisis) begins with some considerations. Why do collectors purchase fine art? To some collectors, the art they buy is a way to access new knowledge: they might not love reading or being lectured, but they do love to be inspired about new ideas by the art they choose. Some collectors like to display their entire collection in their home, whereas other collectors like to keep it in storage and enjoy single pieces from time to time. This way of experiencing art is the opposite of the muscular collections of tycoons and corporate leaders. The intimacy of a very particular painting, selected because of passion and for personal reasons, experienced day after day in the privacy of your home, is a mysterious, almost mystical experience. However, almost every collector will want to have their own best works travel to museums and shows, for reasons of personal pride in being associated with those institutions, and for the economic advantage in seeing the valuation of their investment grow.

Considering the current state of the global economy, the question arises: Will private art collecting turn into an unaffordable luxury, as many other hobbies and pursuits already have? In these circumstances it will be interesting to see if and how collectors will keep on collecting. The general impression so far is that those who entered the contemporary market purely as a way to show off, or with the desire to make an investment with short-term returns, might leave the stage. They will be missed greatly by auction houses, secondary dealers (who are mostly just sales-driven), and other commercially oriented agents in the sector. However, their departure will help others, those who will not give up their genuine passion, to enjoy a commercial environment driven less by finance and more by the quality of the artworks.

Let us take a step back and ask ourselves: How does an average person with a reasonable degree of wealth become a collector? Let us imagine an art lover starting from their first purchase. If we drew a generic sketch of their progression as a collector, it may look like this:

- Our collector-to-be will initially go to a major gallery with an established name and reputation. The collector will buy a relatively affordable work by a major artist, in order to be reassured that this investment and this sign of taste is valuable in social terms (for example when guests will see the work in the collector's home). To marketers, this moment should offer the opportunity to understand the dynamics of social self-appraisal and the natural threshold of newcomers to their own markets.

- As a second step, our new collector might visit a local or national art fair, and expand their contacts in terms of galleries, dealers, and artists. They could sign up to the museum friends association of a couple of good museums within reasonable reach of their home base. The new collector will become more informed about, and get invited to, openings and events. They will meet fellow collectors. The new collector's taste in purchasing will begin to be commented on and discussed in the close circles of the art market. This is probably the crucial moment when the nature of their consumption of fine arts switches from a social signal to a more personal, intimate, sophisticated selection. Brand managers might see here the evolution of a customer from generic interest in the category to the levels of knowledge and competence that make that customer an informal opinion leader.

• Our newly "mature" collector will select and choose a primary gallery and leading dealers to work with. This will most likely be a mix of prominent national and international galleries, offering the collector the opportunity to build up their own selection of works, to follow their own intuition in a discussion and to engage in a relationship that is both mercantile and intellectual. Collectors at this stage of maturity use the expression "I work with this gallery," not the more trivial: "I purchase art from this gallery." The collector may become interested in auctions, in order to, for example, acquire a specific work by a specific artist. Intimacy and a mature discussion with an expert customer are what corporate marketers should naturally aim for, at least if they are serious about community building, and realistic about working with influencers and key account customers on the longer term.

The above might look like a pretty common path, and we might limit ourselves to tracking it in pure economist terms. Let's change the approach, and instead think about the deeper dynamics in the relationship of an art collector to their passion. The major lesson to be drawn from the world of collections is simple: collectors do not purchase purely for rational reasons, and they do not manage their collections on the basis of pure social signaling or irrational passion either. The complexity of each collector, the fine texture of their personality and how this is reflected in selecting purchases, is what makes this field of study a golden learning opportunity for marketers. Such complexity should remind every marketer, every researcher, and every business manager that each customer is a very complex, fascinatingly put together, and unique human being. I hope every marketer and brand manager will rediscover such a taste for uniqueness and truly re-engage in studying individual customers' hearts and minds.

Finding 20: From personal engagement to advocating culture

How private collectors turn into museum patrons

In this chapter we began with how art communities can move from informal gatherings of future talents to become established movements with a media impact and commercial success. Then we looked at the journey of corporate players from successful economic actors to cultural agents, through their investments in art collections. Then, we focused on the collectors themselves, to understand their motives. Now we turn our attention to how an individual's passion for fine arts might lead to institutional sponsorship. Within this "finding" we will understand patronage of the arts, and its socio-cultural background. Here, we move our focus from business entities supporting culture, such as corporations, to private individuals, focusing on how very wealthy individuals can become engaged in the sponsorship of contemporary fine arts.

As we saw with Arte Povera and conceptual art, contemporary high culture might not be palatable for the rest of us. The natural evolution of the fine arts disciplines has not contributed to their understanding by larger audiences. If you consider wealthy individuals who are not in a major city, the challenge for contemporary art institutions in need of financial support is to attract them. It must be remembered that for these individuals their time is money. This challenge can be met thanks to dedicated programs of fine arts education and privileged access for (prospective) sponsors. Such assets and offers are however not nearly enough. Cultural sector managers need to fully understand the ways in which suitably wealthy people think, and create platforms that function accordingly. This is the crucial job of museum directors, and one of the most critical business sides of their work. The manager of an art institution that aims to be financially viable because of sponsorship must study the behavior and habits that entrepreneurs and business leaders adopt in the local area, in order to structure an offer accordingly. Here, there is nothing that can beat an insight in how hands get shaken and deals get closed in the local field of operations of the museum at hand. Within each region of this world and its societal circles, there will be practices and rituals of informal welfare, of art patronage, and of business networking. Regional networks are vital to the sponsorship of any cultural institution. But rather than talking theory, let's look at an example of how this process works.

Noord Brabant is a rich region in the south of the Netherlands. It has enjoyed a wealth explosion, and since the mid-1940s has had an international museum of contemporary art, the Van Abbemuseum. As we discussed earlier, this museum has developed and maintained a rigorous policy of experimental exhibitions and cutting-edge acquisitions, and has an outstanding collection at its center. As with all Dutch public museums, the Van Abbemuseum benefits from public funding, but this is not enough to secure the viability of its very ambitious projects and programs. Additionally, politically charged shows and highly advanced policies do not necessarily help to gain the favor of wealthy businesspeople, who are caught between pride for their city's major cultural institution, and the potential backlash of being exposed to forms of art they barely feel comfortable experiencing.

This is where the Stichting Promoters of the Van Abbemuseum, a private not-for-profit Dutch foundation, comes into play. Members of this association are de facto sponsors, with an entry fee that enables specific museum projects to take place. But there is more. There is a profound connection with those informal networks and rituals that "make" a business region. Led by chair Philip van den Hurk, a successful entrepreneur and major regional collector, and endorsed by the museum at its highest level, this association works according to the business practices that made the city prosperous during the last century:

- The objective of this foundation is not to please the sponsors with some sort of passive event every now and then: Promotors are invited to join the discussions and to contribute to specific acquisitions on the basis of a profound discussion, where they are introduced and accompanied on *their* path of knowledge as much as they financially support the museum on *its* path of growth.

- The promotors are not just a bunch of insular individuals with some sort of interest for art. They are managed like a club and they are nurtured to become one of the important networks that run the local economy in terms of business relationships, shadow welfare, and informal lobbying. This dimension is where the chair makes the connection between the museum (an international body with foreign management) and the local society (a tightly structured and informally textured patchwork of highly interactive individuals who connect in specific informal ways).

- There is of course a specific program, with guest lecturers, special evenings at the museum and ad hoc trips to art capitals or major events such as the Venice Biennale. This aspect, if not the most subtle, is without doubt the most attractive at a superficial level, providing a slightly playful and basically educational element, and addressing a social function too.

The Stichting Promotors is a fairly standard operation in the Dutch cultural scene. It is however different in that it manages to connect potential donors to the contemporary art grammars and the complexities which museum policies generate. The integrity of the institution (in its ever-changing, dynamic nature) is always central to the efforts of the promoters. This is a journey of seduction and socialization, where the museum opens itself without changing itself for its promoters. It seems logical that this specific entity is chaired by a passionate collector, who combines commercial talent with an admirable personal passion and an instinct for the best among fine arts.

Wrapping up: from fine arts communities and collectors to the new role of integrity

The art world had a long phase when auctions and fairs formed the market rather than any content-driven research gallery. In this context, major traders such as Larry Gagosian or Charles Saatchi determined to a large extent the trends in fine arts, with a strong eye for commercial success. It seems reasonable to think that all of this will greatly change as a result of the economic downturn. Looking at the social role of galleries in the life of their own community, we might already find more inspiring practices than Gagosian's commercial drive. There are new ways to reconnect art with society – in Tokyo, the Nakaochiai Gallery (created and managed by Julia Barnes and Clint Taniguchi) is an example of this. It is located in the popular heart of Shinjuku, in the anonymous area by the Nakai underground station. The gallery had to face the challenge of connecting to a neighborhood of local communities: everyday folk who are not immediately drawn to contemporary art. Thanks to its open door policy and to projects that were designed to be taken on board and embedded in the community, this gallery managed to change its perception by locals from alien presence into one of the vibrating pulses of the district. Engaging in discussion with the people and creating events where artists and the arts connect to the people, Nakaochiai shows a different view of working – one that could become a blueprint for 10, 100, and 1,000 galleries and companies in the world.

With such a positive example of connecting community to commerce on our minds, it is time to end this chapter. We discovered art movements that were born and grew in conflict with consumerism. It seems paradoxical to use them as best practice of how some exceptionally talented marketers, such as Siegelaub, and some dealers with unique integrity, such as Tucci Russo, translated the ideas and visions of such art, and transformed them into commercial success. From this challenging starting point, we can list the following highlights:

- Fine arts communities are a form of organization that requires a marketing exploitation approach based on a high degree of integrity. Commercial usage of upcoming new trends can only happen with the utmost respect by the marketer, who ultimately has to be part of the circle of trust that enables the bridging between ideas and markets.

- The conversion of fine arts movements into commercial success does not happen by means of pure exploitation. Even in the face of the Charles Saatchis of this world, commercial exploitation of signature artists is nothing compared with the solid, respectful, and long-term work of organizing events, initiating debate and feeding the cultural process with genuine input. The real thing will last a long time; the current "fast food art" will become just another rapidly disappearing financial investment for all parties, from corporate to private collector, from dealer to museum.

- Regarding corporate collections, the next developments are clear. Respecting the integrity of the curatorial role and – possibly – finding new ways to change the way corporations relate to the cultural sector are the key priorities in this dynamic side of the art market.

- It is vital to reassert how crucial individuals are in the collecting process, with their passion, intuition and emotional drive. An attention to individuality needs to be regained by marketers in general.

- As much as the integrity of art institutions must be preserved, the actual role of sponsors and "friends of museum X" is fundamental keeping the feasibility of programs and projects within reasonable reach and also ensuring a primary circle of socialization for even the most audacious ideas on display. Here, the ways of working of regional business networks are a key reference for all stakeholders.

There are reasons to be optimistic. TEFAF 2009 (the Maastricht fair of fine arts, from classic antiquaries to modern design) recorded a record number of applications, and demonstrated a great degree of commercial vitality. Around the same time, in the darkest month of the global economy for decades, the exceptional sale prices achieved by Yves Saint Laurent's art collection, auctioned in Paris, were rare good news. I have already anticipated how it is the nature of these times of crisis that the major leaders in each industry see a benefit in terms of the perceived value of their solid roots. There might be no specific reason for hope, but it is always a good sign to see the recognition by the market of the best propositions. It confirms that quality and excellence are the best ways to successfully tackle the crisis. Rediscovering the bonding of people with people, and of people with artifacts, will be key to the success of tomorrow's business leaders and marketers.

Understanding the obsessions in a collector's mind

A collector does not just buy objects: a collector selects, edits, curates, anticipates, dreams of their collection, how it is and how it will grow. Many theoretical studies have been performed about collectors and collections. Psychology and psychoanalysis point to the deeper meaning of this practice, economists discuss it from the viewpoint of value creation, and art critics review it from aesthetic and philological viewpoints. It is not the purpose of this book to go deeper into this kind of analysis: our aim is to highlight just how collectors make the markets by making their collections, and what can be learnt from this for the purpose of more general business practice.

We find the best clarity in the extremes, and then re-emerge with universally valid considerations and conclusions. For this case study we have chosen to refer to a fictional case, rather than any scientific study. This fictional case introduces the psychology of the collector in its richest form: a literary character created by former Sotheby's director turned journalist and novelist, Bruce Chatwin. Kaspar Utz, the protagonist of Chatwin's final novel, *Utz,* is a citizen of Prague and the son of a wealthy family. Across the dramatic history of his city in the twentieth century, he experiences, like all his fellow countrymen, the invasion by German armies and then the occupation by Soviet Union troops. He will see the pendulum of history swing from one dictatorship to the other, and he will experience tragedies as only Eastern Europeans have. However, Utz shuts himself out of "grand history," he shuts himself out of real love, he shuts himself out of anything not specifically pertaining his one passion – his Meissen porcelain collection of antique figurines.

You could say that Utz (a man whose existence is devoted to the fine art of collecting) is pathologically alienated from reality. In actuality, he is a perfect, extreme type of collector. His passion is the El Dorado of any dealer or gallerist. The fact that he chooses to have his collection destroyed at his death, rather than it being absorbed into a state museum run by the communists, only adds a slightly anarchist twist of color to the character. Chatwin describes the über-collector, and in doing so, demonstrates a number of mental processes that are behind a passion for collecting that becomes obsession. Although you might not like Utz for his complete lack of social sensibility, it seems important to refer to him as the ultimate possibility of a complete drive towards collecting. Or, if we were to vulgarize this rather sophisticated story, think of Utz as your possible best customer if you were the incumbent dealer in the market of Meissen porcelain figurines for collectors. Think of your own enterprise as servicing somebody with the levels of passion and commitment that Utz demonstrates across this short novel. And then, think of how such a mercantile connection would be a match made in Heaven, and paradise on earth for any marketer.

CHAPTER 6

Between "ideology" and "space:" the museum experience

This chapter deals with the way contemporary museums are being designed, experienced, and marketed. We explore two fundamental dimensions of museums:

- "ideology," or the expression of a museum's essence – its soul

- "space," or the physical presence of museum buildings and premises in cities, and their design principles.

The rise of museums as contemporary, nonreligious cathedrals reached a peak with Frank Gehry's 1997 design of the Guggenheim in Bilbao. That building represented a chance of renaissance for the Basque capital, and the beginning of a new era of museum-driven cultural tourism through regional airports. After such a success, a number of cities in the world adopted the same approach: a great architect, an outstanding building, and an aggressive marketing drive. However, with a few exceptions, the outcome of this gold rush to replicate the Bilbao model led to disappointing results. The somehow exceptional Bilbao case of iconic city branding through museums is one of the most memorable and notable examples of how museums took center stage in becoming relevant well beyond their traditional purposes. We explore leading ideas behind these institutions, rising trends in their complex universe, and how business leaders can benefit from them.

The business question behind this chapter

How can contemporary museums inspire strategy and marketing visions?

An introduction to the findings in this chapter

In the 1900s, it was not unusual to read in articles and books that the contemporary museum is the cultural equivalent of the cathedral in the Middle Ages. Of course, museums as we know them today come from a different angle and with different purposes than religious buildings. Our contemporary museums are the long-term evolution of the private collection room of the Renaissance nobility in Florence, and especially of the Louvre in France. In 1793, the Louvre was opened to the public on the first anniversary of the French revolution. After its opening to the masses, the Louvre flourished. In a parallel line of evolution, the British Museum (created in 1759 for different reasons) developed into a public museum. Deliberately managed as ideological centers during the 1800s, national museums became part of the natural portfolio of rising nation-states, and an important feature in the social construction of specific identities of regional culture. What we discovered in Chapter 4 about the history and the role of Amsterdam's Rijksmuseum holds true as a general principle.

One last example of the length of the ideological journey of museums from their origins to today: The idea of the classic museum was to some extent anticipated by the secluded *Wunderkammer* (cabinets of wonder) of wealthy Renaissance aristocratic collectors. This was the closed and private environment where the whole world was supposedly for the privilege of the wealthy elite: The known universe was created exclusively for just one patron and his close circles of friends and family.

It is easy to appreciate how distant this notion is from our contemporary museums, which mostly have an educational public function as their mission. A visit to the design section of the Museum of Modern Art (MoMA) on an ordinary New York City Sunday morning will entail strolling around groups of kids with their parents, guided by a museum educator who fully engages with them, inspiring their curiosity and their thinking about what design is, and why design is specific in its functions and manifestations. Children and young adults walk around the rooms with their pencils and notebooks, making notes and sketches. The same can be experienced in any major or minor museum in the world. This field of analysis is immense, and a matching literature is there to explore it.

Generally, it is fair to say that the organizational machine of museums as we know them today was engineered in the 1930s by Alfred H. Barr, Jr., and his team at MoMA in New York City, world capital of fine arts since the second half of the 1900s. This museum is still the template for most contemporary art, modern art, and visual culture (design, architecture) museums in the world. We discover more about the historical roots of this specific institution in Chapter 7 on design.

Museums matter today more than ever. To further our understanding of the reasons that they do, we examine different levels and categories of analysis. As the title of this chapter suggests, we look at two extremes as our reference points:

- "ideology" as the final outcome of functions performed by museums and the way they are translated into signs and symbols, for example, identity design systems

- "space" as the design of the experience that determines what a museum is, in the physical terms of bricks and walls and those large windows that enable people to visually explore beauty and self-reflection, as they did in the cathedrals and churches of the past.

Based on these two main categories, we explore how museums with their rich history and their challenging future can inspire and improve our business practices. We do so through our five "findings" for this chapter:

- *Explore the different facets of who you are, to fully understand the meaning of your actions*: We look here at different descriptions of the role and function of museums, with "soul searching" as the key concept in this environment.

- *Conceive your corporate identity systems to always truly express your cultural DNA and "soul:"* This "finding" focuses on specific examples of how museums are branded from the viewpoint of identity design and art direction.

- *How buildings express what you are*: Engage in a new debate with the physical infrastructure that your enterprise inhabits: mixing ideology and space we explore issues of urban architecture and interior design.

- *Express your soul by connecting to your urban environment*: Connecting buildings and cities to reconnect enterprise with society.

- *Express your soul by connecting culture and commerce*: Connect museums and business. The ways business leaders can establish innovative connections between museums and commerce. Finding the way enterprises may profit from culture, and vice versa.

As we discovered in Chapter 4, Damien Hirst's show at the Rijksmuseum in November 2008 demonstrated how even a classic museum of international stature can prove to be very dynamic in terms of both curatorial experimentation and brand marketing. We now review a number of cases further describing how museums offer business leaders and brand marketers unique discoveries in terms of strategic vision, design management, and operation.

Finding 21: Learn who you are by searching your soul

Explore the deeper meaning of your internal culture

In this "finding" we focus on the ideological dimensions of museums, and the different possible personalities that a museum might embody with its policy. We explore a few relevant examples of what this kind of institution does, and how it can vary in its description. This will remind us practitioners of the creative industry and of the worlds of marketing and branding, how crucial it is to know oneself first of all, in order always to keep a clear focus on the meaning and the essence of what you do. To start such a complex topic, we explore three very different ways to look at museums and the approaches that govern them, in general terms:

- *The "philological" approach*: Museums in the major league have a structural, almost implied function: that is, preservation of heritage and history. This is the academic, institutional approach, which we define as "philological." According to this vision of museums, what matters is the taxonomic display of objects and artifacts with a historical focus. Within this concept, museums are scientific bodies with the duty to objectively provide the cultural and aesthetic history of the manifestations of an age. Any traditional public museum with an informative, rigorous and neutral display of paintings and objects will represent this approach.

- *The "experiential" approach*: At the opposite extreme, we find the notion of "experiential" museums that border "entertainment." Here, the display of objects within the museum space has the purpose of enticing and appealing to the viewer, sometimes leading to luxury retail and high-end leisure architecture solutions.

- *The "radical policy" approach*: This museum policy is perhaps even more extreme a source of ideological tension. The purpose of a number of outstanding contemporary art museums is to pursue a radical path. This means organizing the most outstanding contemporary art shows, and bringing them to the public with their political and aesthetic strengths, without any fear of an adverse reaction. Of course, this is more a curatorial approach and it might therefore be combined with one of the two more fundamental approaches as illustrated above.

What connections can be made between these approaches to museum management and our everyday world of business? With the "philological" approach, we can draw some parallels with the reality of corporate enterprise. It is of course important to ask yourself, as a business leader, whether your organization maintains an accurate philology within its portfolio in terms of historical consistency. Examples thereof might be the recent five years of design experimentation by FIAT Group Automobiles, with the aim of reviving the spectrum of signs and symbols that made the company successful in the past. This was the starting point for FIAT to attract President Obama and the United States, with its stylish, environmentally friendly city cars, like the world-famed 500. This is an example of a philological approach to product portfolio management in terms of heritage design, given its explicit connection to the 1950s model bearing the same name.

Moving to the experiential design level, there is a decade or more of business literature to tell us how crucial such a notion of brands is. Here – at the other end of the spectrum to philology – we can recall the sophisticated evolutions of luxury retail and the rise of experience centers working on equivalent lines to museums, such as the Kristallwelten by Swarovski in Wattens, Austria. On the other hand, returning to experiential museums, in recent years a number of them have been managed with an eye on marketing. The pursuit of commercial success has resulted in the bastardization of policies and programs with the result verging on kitsch. An example of this is the Groninger Museum in the Netherlands, with its productions of show-biz-styled events like the "Fatale Vrouwen," opened by guest star Joan Collins. These museums represent an example of how commercial prioritization can result in the dilution of the curatorial policy in the first place, then of the brand, and an ultimate loss of credibility.

Concerning the third approach of "radical policies," we return to the Van Abbemuseum in Eindhoven and to institutions with an equivalent policy of radical research such as the New Museum in New York. Companies such as Benetton and Diesel have used this approach in the past, and more lifestyle leaders will do so in the next decade, when radical and critical thinking will be the driver of socio-cultural trends.

In conclusion, the challenge for business leaders is to reevaluate where the core of their organizations truly lies. Vital questions for business leaders and brand marketers include:

- Is the history of your company philologically placed in its current offering?

- Is the experiential dimension of your brand aligned with your DNA?

- Can your brand benefit from explicitly addressing more radical viewpoints, especially with reference to burning societal matters?

The above questions could appear to be removed from everyday business practice. To engage in this kind of analysis could be a big leap. However, this was the case for museums in terms of their function at each stage of their existence. The role of a "national museum" was fixed, and very different from the educational role of museums today. Museums used to be just "philological:" they rapidly turned to experience-driven design strategies. It was only a matter of time: Radical change just happened in terms of ideology as demonstrated by policies over time. It is a matter of truly searching your soul, as profoundly as possible, and then emerging with innovative ideas to redefine the way your business is done. How will this rebirth be made tangible and visible? Once this journey of self-discovery is performed, design is the best discipline to represent its outcome visually.

Finding 22: Create your brand identity to express your personality flexibly

Conceiving corporate identity systems that reflect the essence of who you are

In our first "finding" in this chapter we established the crucial role of "soul searching." It is time to move our focus from "soul" to "visual presence," or from vision to design. The power of design is to render ideology tangible to the eye of the public; the power of designers is to manage to influence such ideological levels because of their creative edge. Our fellow traveler for the initial stretch of this second "finding" managed to do both in its past: it is Thonik, the design agency we mentioned in Chapter 4. We begin our journey from the Centraal Museum of Utrecht, one of Thonik's 1990s major projects of soul searching by means of identity design.

Founded in 1838, the Centraal Museum is one of the oldest municipal museums in the Netherlands. In what could be seen as a reflection of the Dutch traditional way of life, variety is the nature of the Centraal Museum. The Netherlands have grown in prosperity and fortune thanks to a system of pillars, or *zuilen* in Dutch. Society was traditionally "divided to unite it" in different groups. Each group was described as one pillar. The constructive interaction among them would bring the consensus with diversity that so much benefited the country in its liberal and tolerant past. Perhaps not explicitly designed with this purpose in mind but nevertheless modeled on this cultural imprint, the Centraal Museum in Utrecht is divided into five completely different departments:

- Old Masters

- modern art

- design

- fashion

- local history.

To add to the organizational profile and to the body of this institution, the museum also features among its assets:

- a permanent collection of the works of world-famed children's cartoonist Dick Bruna

- the world's largest selection of works by modern master of design, Gerrit Rietveld

- a brilliant design by Droog Design that has in recent years redefined the museum interior, beginning with the vibrant restaurant by Richard Hutten.

It might seem challenging enough to handle such a level of interdependent complexity: it must then be added that the five departments were notorious before the mid-1990s for curatorial issues. How can identity design express a sense of coherence, consistency, and purpose for such a complex organism? Thonik used its cutting-edge creativity and defined an identity system flexible enough to represent the nature of the Centraal Museum and its pillars, or departments. It is based on a simple X-shaped structure defined by five Cs, four at the extremes of each arm of a St Andrew's cross, with a fifth at its center. Here, each of the five Cs represents one of the five museum departments. Organizational and curatorial complexities find in this living logo a visual organizing principle to tie the company together as one brand.

This system can be used flexibly in various modes of execution, for specific campaigns. Its lightness makes it fit with the rich nature of diverse visuals within the different programs of the Centraal Museum. This is an identity system that does not force one artificial logo on the entire museum: It instead leads to flexible coherence and multiple configurations where variety and diversity are preserved.

The question raised by the Centraal Museum and its identity is: How does your corporate identity truly represent who you really are, with the uttermost flexibility to change and adapt in order to always do so? The main point of this "finding" is that identity design and brand management should be tools in the hands of brand leaders who search, define, and manage the brand personality. This is not the standard case in the corporate world. In classic corporate identity terms, consistency and coherence appeared as the only bottom-line value. Manuals and textbooks rule in a situation where means have mostly become the purpose: The practice of identity design as we knew it for the last decades has been turned upside down. The world of fine arts museums contradicts this sacred cow, and the Centraal Museum is not the only successful example of this.

It is time to expand our analysis from one case to another, that has taken place over many years. To do so, we return to Van Abbemuseum. We discover the particular relationship between its directors and its corporate identity as revealed in brochures, catalogues, and other graphic representations. From a curatorial policy viewpoint, the Van Abbemuseum has always been ahead of its time, anticipating the next art movements, the next countries where leading art will come from, and the next societal developments. In branding, this line of conduct might be described as "brand behavior." At the same time, examining the connection between the different directors and the different graphic design guidelines in its history, we can identify flexibility in its brand design over the long term:

- From 1946 to the early 1960s, director Edy de Wilde worked with Dutch graphic design legend Wim Crouwel, for whom the museum brand and its consistency stood above all. Clarity, coherence, and rigor were Crouwel's drivers during his years, starting after the Second World War and ending in 1963.

- With Jan van Toorn, chosen by director Jean Leering to work with him between 1964 and 1973, each artist and each show received a special visual treatment. These were the years of experimentation and of challenging, if not the ideological collapse, of traditional top-down authority and structures. With elegant execution, Van Toorn's art direction followed the same ideological lines and therefore resulted in an almost anarchist overall design presence.

- The director from 1975 to 1987, Rudi Fuchs, effected a 180-degree change in policy, from collectivity-driven innovation to the restoration of the autonomous art artifact as the core of the museum. The art direction by Walter Nikkels might seem visually "frozen" compared with the exuberance of Van Toorn's works. It is however in line with (and perhaps even anticipated) 1980s postmodern design by the likes of Peter Saville for new wave bands such as Joy Division.

We stop here, because as much as we might continue with the analysis the song would remain the same. Under each director, the art direction of the Van Abbemuseum greatly changed. Under each art director, from Crouwel to Van Toorn to Nikkels, the corporate identity represented a best in class outcome in terms of standing as a sign of the times. This historical pattern leads to a true brand marketing paradox. Looking at the complete portfolio, we could say that the Van Abbemuseum corporate identity has not been well managed across the decades: Not many business enterprises would change their visual presence so often, in such radical fashion. It is instead the conclusion of this "finding" that such flexibility, such radicalism, and such willingness to dare are good. We believe that such features are good in general, and we definitively believe that they were excellent for the Van Abbemuseum. The outcome of decades of art direction for this experimental institution shows a coherent line of inconsistency from decade to decade. This reflects the nature of this institution to anticipate the spirit of the times. Over time, at the Van Abbemuseum, brand identity design followed the evolutions of its ideology and of its personality. I believe this is brand design at its best.

Finding 23: How buildings express what you are

Know yourself from the very essence of your bricks and windows

A museum building is one of the most crucial embodiments of its ideology. It determines its presence in urban space. It is the enabler of its experience by visitors. Within this "finding," we move the focus of our analysis of museums from ideas to "space," or the architectural presence of museums as buildings, both interiors and contemporary metropolitan icons. In the first instance, we look at some insights related to the physical dimension of museums, with particular focus on their interiors. We start with four considerations:

- *Museums are complex architectural programs*: There is more to a museum building than just exhibition spaces. There are social spaces, retail, restaurant, versatile spaces to convert flexibly for events and meetings, administration offices, and more. Museums are multi-purpose programs, to be addressed holistically by means of careful planning of the visual, sound, and tactile experiences.

- *The museum experience is a multiple one*: Visitors want to have a two-way experience with art: They want to be in a state of mental isolation in front of masterpieces but also to feel part of a multitude, of the social construct of museum visitors. They meditate, memorize, observe, inform themselves, amuse themselves, all at once, all in the same exhibition rooms, alone and all together at the same time. Once again, careful architectural planning is a must to ensure the right degree of intensity versus relaxation across this journey of wonders.

- *The museum entrance hall is key to the overall quality of experience*: Here, we will find contradicting mixes of needs and demands, from protection from the rain if queuing to proper storage for your outdoor clothes, briefcases, and bags; from an efficient ticketing process to the availability of information at any given moment;

- *Going into the museum is an emotional rite of passage*: Beyond functional and utilitarian considerations, this moment represents the passage from the "sky of God" outside to the museum shelter inside: almost a moment of spiritual nature, signaling an almost mystical threshold.

If we think of the implications of these bullet points in terms of corporate interiors and branded architecture in general, the following questions arise for business leaders and brand marketers:

- To what extent do your corporate architecture and retail programs put people at their center?

- To what extent is your corporate architecture and branded retail designed to provide people with multiple stages of experience, beyond the sterile efficiency of globally deployed formats?

- To what extent do your corporate design strategies for buildings and branded retail take into account the specific experience of entering the space from the outside, valuing it as a key brand connection?

An overall question for business leaders and brand marketers is: What are the social and emotional connections that your buildings inspire, whether they are corporate headquarters or brand retail flagships? Especially from the perspective of history and heritage, we might say that such a socioemotional dimension has not always been truly exploited in the last decade or so. Caught between economic boom and the downturn, corporations first opted to build new architectural icons of spectacular presence, but without historical bonding with people. They then panicked and abandoned entire complexes to save costs. Leading on from this vital question, we continue our analysis of the Van Abbemuseum history. Now, after reviewing its policies, its sponsoring foundation and its art direction over time, we focus on its three buildings.

The Old Building was built after a major donation by the tobacco industrialist Henri van Abbe. The project was assigned to architect A. J. Kropholler, who between 1934 and 1936 created a building in line with the Delftse School principles, strongly influenced by the ethical, spiritual, and aesthetic trends of his time. The 40 x 40 square meter design was realized in red brick with a clock tower. This remained the state of play of the Van Abbemuseum until 1995, when the museum was shut to start work on a new building. The exhibition programs were temporarily relocated in an old industrial loft space in the heart of Eindhoven, under the name Van Abbemuseum Entr'acte.

This is the second building, where the museum was for seven years under the direction of Jan Debbaut, with important shows by Atelier van Lieshout, Douglas Gordon, and Marijke van Warmerdam. In 2002, along with the New Building, an important extension to the Old Building was finalized and opened to the public. The New Building, with its iconic exterior and its highly fragmented interiors, was designed by Dutch architect Abel Cahen. Here are some reflections on these three different venues:

- The Old Building became so beloved by the people of Eindhoven that the idea to replace it completely with a new one was strongly opposed. This led to it being reclassified as a national monument to be preserved: Its interiors are still the same as they were in 1936 when it opened.

- The Entr'acte building, today housing the young and regional Temporary Art Center (TAC) was supposed to host the Van Abbemuseum for a two-year period only. In its seven years, the flexible interiors offered talented curators and artists a blank canvas to create and curate highly innovative shows.

- The New Building, with its almost labyrinthine structure, was the source of inspiration for the new "Plug In" curatorial policy: this is a particular way to present selections of the unique Van Abbemuseum collection in combinations and settings defined by "invitation only" special guest curators. The "difficult" interiors led to substantial innovation, opening the museum to a dialog with the elite of the fine arts, for the best display of the museum's important collection.

These conclusions open three direct and challenging questions for business leaders, related to the "interiors" of their business:

- *To what extent is the history of your corporate buildings and retail spaces used and appreciated, in terms of its emotional connection to employee and customers?* Here, the current economic downturn might help slow down the "race to new iconic towers" that in the 1990s and 2000s saw a number of corporations build new, anonymous skyscrapers, instead of truly nurturing their history and soul, as the 1936 Van Abbemuseum building does.

- *To what extent is your organizational layout of interior spaces innovative and supportive of creative experimentation like the Entr'acte building was?* In Chapter 3 we discussed how desk space policies and other Orwellian approaches to interior design might act as a killer to the output of creative industry teams. The question now is how to reverse this situation, and create spaces that do work to support and encourage creativity.

- *To what extent are the interiors of your facilities truly leveraged for what they are, in order to extract from the venue new potential ways of working, as in the case of the Plug In curatorial policy?* The deployment of standard modular office and retail solutions has been one of the myths of standardized globalization. Why not go back to the contextual power of more local concepts of the spatial organization of work?

The conclusion of this "finding" is simple. Leading museums have been developed to the highest level of sophistication in terms of both urban presence and interior design. Their relationship with cities is often based on a charismatic bonding between their buildings and people. Analyzing museums as a source of inspiration (and aspiration) for corporate architecture and branded retail could be a great help. Ultimately, the objective for business leaders and brand marketers should be to prevent the sterile alienation of corporate architecture and/or the claustrophobic feeling of so many branded retail flagships, headquarters halls, and open-plan buildings. All the passion and all the magic of the museum experience could be transferred to the actual physical body of your company: would it not make it a better company, by design?

Finding 24: Express your personality by connecting to your urban environment

Connecting buildings and cities to reconnect enterprise to society

In Finding 23 we demonstrated the firm emotional connection between the Van Abbemuseum and the citizens of Eindhoven. Continuing this theme, we now move on to discover further connections between the museum and the city. There is a "before" and an "after" regarding urban museums. As I anticipated, this dividing line is 1997, when Frank Gehry completed the Guggenheim in Bilbao. In pure architectural terms, the building is extraordinary, and a milestone in the use of computer-aided design. In terms of the relationship between museums, cities, and architectural icons, Gehry's work generated a wave of investments in every continent, with the sole purpose of replicating the same strategy of city marketing. This led to a proliferation of urban icons in the world, ultimately diminishing their overall impact in global differentiation terms. It cannot be stressed enough that each architectural icon is like a kid shouting on a square, and a square full of shouting kids results in nothing but unbearable noise. It should also be said that this drive towards outstanding museum architecture has resulted in the paradox that such iconic buildings steal the show from the art they are supposed to protect, shelter, and exhibit. The result is that art becomes ancillary to the architecture that should serve it. This paradox about the function of museums versus artworks leads us back to the ideological dimension of these institutions, connecting architecture with strategy at the highest levels.

In rational terms, the main purpose of a museum building should be to protect and serve its art. What are the other options in terms of the urban planning and city branding function of museums? We address this key question with two contrasting examples that connect both dimensions of ideology and space. At the time of writing, the municipality and the public opinion of Amsterdam were immersed in a dispute over the strategic choices to be made in order to regenerate the south side of the city, or *Zuidas*. Two parties were engaged in a debate that underlined two opposite ideologies of what the museum is and what it should do, at both curatorial and city branding policy levels.

On one hand, leading authorities such as Martin Sanders (former director of the Concertgebouw) and Wim Pijbes (current artistic director of the Rijksmuseum) indicated their belief that an iconic building in the same fashion as the Grand Palais of Paris should be designed and built to attract the international tourist masses to the south side of Amsterdam. Here, selections from the grand collections of the Rijksmuseum, of the Van Gogh Museum and of other Amsterdam institutions should be used to create blockbuster shows of great international magnetism. On the other hand, design innovators such as Platform 21's director, Joanna van der Zanden, have been working for several years on an alternative design museum format. Her thinking leads away from an overwhelmingly alien grand palais in a secondary area of town. Van der Zanden pushed more towards innovative, radical, open-ended formats of interaction with the audience, and of cultural experimentation. This thinking would obviously demand a completely different architectural approach. These two opposite ideologies of museums also represent two polarities in terms of spatial organization. Here, our analysis of the contemporary status of museums finds a point of synergy between ideology and space: the closed philological experiential machine for a grand palais, versus the open, socially driven form of new radical formats. The natural question that follows is: What will be the next evolution of museums?

What museums deserve in the future is a different strategy than the "iconic building" approach described above. What about a strategy that puts the true genuineness of the art process back at the center, not forgetting the creation of a strong emotional bond with the city? Architects working on future museums should be less like late 1800s "salon painters," working for any neutral white gallery space or home room, and more like early 1900s muralists, who strictly adhered to the *genius loci* of place and environment. The main challenge will be to move from the rush to new icons to the new socially driven design of places for people to be. The most natural way to achieve this in advanced economies and in emerging markets is the regeneration of brownfield land and industrial heritage through the creation of communal cultural spaces. Of course, not everything is that easy – fundamental details such as the climatic control of environments present huge cost challenges. Nevertheless, this approach might work very well. It led, for example, in Turin, to the creation of an art exhibition space in the otherwise oppressive architecture of the old prison building, Le Nuove, with a very positive outcome. This is just one of the countless examples of positive regeneration in the world, from Western Europe to the art centers of emerging cities such as Beijing and Shanghai.

What kind of inspiration can business leaders derive from these museum-related trends? The immediate questions to be asked are:

- Have you defined your overall strategy when it comes to your buildings and retail?

- Is your enterprise still pursuing old strategies such as the creation of new architectural icons?

- What are your regeneration strategies with respect to your old buildings from redundant production centers in what were our global advanced economies?

This last question should be considered equivalent – in terms of relevance – to the environmental challenges of product recycling and waste management. Of course, these considerations should be applied to retail architecture and interior design as well. It is time for corporations to take responsibility and advantage from their past and current cycles of urban land use and estate management. In a time of transition and crisis, such assumption of responsibility should be put at the heart of what the company does – architecturally speaking as well. The next challenge for business leaders is to immerse themselves and their design leaders in this flow of ideas. The same challenges that apply to the future of museums, such as in Amsterdam's *Zuidas,* are valid for the future of corporate architecture and urban presence as well with the same degree of urgency, if not even more.

Finding 25: Express your personality by connecting culture and commerce

Finding the golden intersection between museums and business

In this last "finding" we move our focus from museums as ideas and/or buildings, to museums as business. We begin with a statement about the nature of any museum, even the most radical and the most avant-garde ones – each of them is ultimately all about sales. Let's verify this assumption with three examples:

- If the museum relies on ticket sales, it will have to excel in the marketing of blockbuster exhibitions to the masses. Think of any private museum in the world, or art institution depending on the public.

- If the museum is based on grants and donations, then it will be all about the sponsors – and how to please them within the integrity and the dignity of the artistic policy in place. This is another form of sales.

- If the museum is based on public funding, the sales task is about the ability to motivate local politicians and authorities to accept sometimes radical programs, or to "sell" the losses from the ticketing of the most avant-garde, difficult-to-digest shows back to the public commissions owning the financial responsibility for the institution.

Museum directors and cultural sector managers are faced every day with the eternal dilemma of every entrepreneur and of any business leader: pursue a vision or sell out? Of course, as in the example of the Rijksmuseum, the balancing act between philology and experience, between education and radicalism, between commerce and conservation is vital. As for any business enterprise in the world, there are plenty of opportunities for museums to earn revenue commercially. Of course, museums have their bookshops and their gift shops. These places are normally disregarded by museum studies scholars and analysts. They would however be interesting for a case study in itself, as they represent the memories in people's homes of the shows they experienced. Here we can speak of the knowledge-creation potential of shopping. Gift shops represent a way to learn, in terms of dispensing information and enabling the preservation of memories related to the museum experience. Think of the particular relationship that some collectors have with their exclusive works of art. Then think of the relationship that millions of people like us have with their art books and museum catalogues. Displayed in living rooms and on coffee tables, these publications purchased as an important part of the museum experience are without any doubt the longest-lasting connection between mass audiences and art, and an important source of commercial revenue to museums.

Another opportunity to get revenue from visitors is the museum café. The vital connection between food and shows is not new, but also not that old: some 25 years ago the presence of a fine restaurant in a museum environment was much less usual than it is today. Airport retail and museum retail experienced parallel paths of growth, with the same target audience of "people in transit with time on their hands." We can mention again the Rijksmuseum pavilion at Amsterdam's Schiphol Airport, a unique mix of art and commerce at the very heart of an intercontinental hub with millions of travelers per year, and a chance to earn important revenue, without compromising the integrity of museum policies.

Even beyond museum gift shops and restaurants, commerce can contribute positively to the life of museums in terms of both financial viability and intellectual stimuli. Is this the case also for radical museums that aim to reconnect to society? This would appear to be quite a challenge. Enter the socially engaged Nederlands Fotomuseum in Rotterdam (the subject of this chapter's case study). The nature and the specific social presence of the Nederlands Fotomuseum provide a point of reference for social integration. You might think that such an institution would not engage in any form of commercial exploitation of its assets, sticking to a sort of liberal idealism or even left-wing vision of its function in society. Nothing could be further from the truth. The dynamism of the Nederlands Fotomuseum in the social field is mirrored by its commercial talent to invent new business models of collaboration with enterprises. Let's look at two examples:

- *Consulting partner*: Companies use the museum's knowledge of photography in order to help their teams formulate their communication strategies. In the past, this museum organized workshops for TNT/PTT, the Dutch postal service provider. This is a company with a corporate communication history featuring several photographers. The Nederlands Fotomuseum advised on the role of photography in TNT communication strategies, at both historical philology and contemporary campaign levels. The museum acts as a de facto strategic content adviser, charging a consulting fee for its services.

- *Content provider*: The collection is made up of 1.3 million images, of which 100,000 are fully digitized. At the time of writing, discussions between the Fotomuseum and the retailer HEMA were under way, with a focus on the business model offered by HEMA for web-based printing. This is a business where HEMA are number two in the Dutch market thanks to an online software application to manage and then print digital images for people's own photo books. Here, the museum developed a specific offering to commercialize its digitized historical images to HEMA's audience of millions. A second example of this kind of business cooperation is the partnership with Hollandse Hoogte, the largest press bureau in the Netherlands. This agency adopted digital formats for imaging within its business model. The Nederlands Fotomuseum does not have a marketing management, or a direct sales capability for the images, reproductions, or digitized files from its collection. Hence, Hollandse Hoogte was empowered to make 20,000 pictures from the museum archives available for sale. Hollandse Hoogte became the shop window of Nederlandse Fotomuseum, with unprecedented mutual commercial benefits.

The expectation at the Nederlands Fotomuseum is there that there will be a large development in both consulting and media-based collaborations. This is also a winning example of museum management strategies to present, diffuse and promote the collection, to make it seen, and to share it outside the walls of the museum building. This way, the museum is wherever an image is being purchased and downloaded. And of course, it is a great case of commercial collaboration between museums and corporations, between culture and commerce.

In conclusion, the key question for business leaders and brand marketers concerning their direct relationship with museums and cultural institutions is: Apart from simple sponsorship such as a logo on the exhibition banner, are there better opportunities for healthy commercial relationships between corporations and culture?

Important new growth territories were created by HEMA, Hollandse Hoogte, and other enterprises willing to enter into a dialog with innovative museums such as the Nederlands Fotomuseum. Who will be the first enterprise to follow these innovative companies, in this new arena of marketing excellence?

Wrapping up: from fine arts museums to a higher sense of purpose

To close this chapter, we return to the museum that defines the standards of all modern and contemporary art museums in the world, MoMA in New York. Its most recent building extension was a minimal design by Japanese architect Yoshio Taniguchi. Here, the collections are awe-inspiring and the shows are staggering. However, the overall experience, with the escalators, screaming children, and the masses of tourists, is more like a supermarket or a busy airport than the austerity of a temple of modern art. Contrast this with the experience of visiting another museum designed by the same architect – the Higashiyama Kaii Gallery at the Nagano Prefectural Shinano Art Museum in Japan. This is not an international museum as MoMA is. The gallery itself is of a very simple, very Japanese design, hosting a local collection of a Japanese master mostly unknown to the outside world.

I visited the city of Nagano and its museum on a spring day, with a magic sunset casting unique shadows on the alpine heart of Japan. There were not many visitors in the gallery, and the light was ideal to experience the reflections of the sun on the water in the artificial pool outside the large windows of the gallery. I felt a real sense of epiphanic peace. Of course, this resulted in unforgettable memories of this museum, and of the works by Higashiyama on display. The point of this story is very simple: you might hire the best architect, and get them to design the best museum in the world. The experience of it however, does not lie purely in the design: it is the magic of a pure moment that is not repeatable – as with many things that are precious in life. Of course, MoMA is and remains the template of a contemporary museum, and the Nagano museum will never come even close to it. However, in the heart of at least one international visitor, the museum in Japan is the ideal, not MoMA, at least from an emotional and spiritual viewpoint. While the world praises the work of Taniguchi in New York, it is important to report the power of his vision in a minor setting like Nagano. This reminds us all that the museum is a place of (self-)discovery, beyond the analysis of business models, commerce, and other similar concerns.

To sum up, in this chapter:

- We observed different ideologies of museums, from scientific institution to experience center. We concluded that the challenge of the business leader is to search and explore the "possible souls" of its own enterprise.

- We reviewed two case histories of different applications of corporate identity programs to museums, and we learnt that a Copernican revolution is possible in terms of designing corporate identity systems not on the basis of stale manuals or old ideologies, but by looking at the very nature of the organizational culture at hand, examining its relationship with the society and environment in which it operates, and expressing a mature degree of experimental flexibility.

- We examined, in Findings 23 and 24, the dynamics of museums as buildings, first with an eye on their interiors and the way they are experienced by visitors and planned by architects, and then the relationship of museums and their urban environment. We saw how questions about future museum formats and locations are relevant to the situation where business leaders investigate the meaning and the function of their corporate buildings, their branded retail, and the organizational models reflected there.

- We explored the way in which commerce and culture interact within the museum environment and demonstrate the need to maintain a highly dynamic view of possible business models to use the assets' potential creatively, just as a socially engaged museum such as the Nederlandse Fotomuseum does.

Going back to the Nagano visit, one could say that in spite of all the iconic museums and the great investments in the cities of the globe to create the next great "marketing machine," the uniqueness of a true moment of delight can only be engineered to a certain extent. Given the work of the right architect and the setting up of the best conditions, just like love, it either happens or not.

Leadership by social integration: The Nederlands Fotomuseum (Rotterdam)

The Nederlands Fotomuseum in Rotterdam is the leading Dutch cultural organization in the field of photography. The museum focuses on the definition of what photography is. It has a conservation approach and storage capabilities that meet the highest level of sophistication in the Netherlands in this medium. This museum is the result of developments in the cultural field, the growth in the recent decades of a greater interest in twentieth century media such as documentary photography and cinema. Its strength lies in a strong policy of acquisition for a collection focused on representing the best photography can offer. It does not follow the standard for photography in the prescribed formats of contemporary art museums. Additionally, its physical location in the Las Palmas city docks complex is important.

What makes the Nederlands Fotomuseum collection important is its policy of acquiring not just single pieces, but entire sets or projects. While most contemporary art museums in the world possess individual photographs by individual artists, this museum uniquely acquires complete projects with little interest in single, notable star pieces. This policy is a tangible manifestation of a very specific ideology of museums. What makes it even more interesting is the relationship it has managed to build within its physical and local environment.

The Las Palmas city docks are the same docks from which ocean liners sailed to New York, decades ago, with the dreams and hopes of European emigrants. The Las Palmas district in the proximity of this museum has 300,000 people of 150 nationalities, and 75 percent of the area is a deprived neighborhood, including the former Rotterdam red light district. When faced with such a social challenge, a museum director can choose either to invest in rebuilding the museum itself or to actively address the social challenges and ultimately contribute to building a better society. This second option was chosen at the Nederlands Fotomuseum. Cultural projects were developed by various parties within the city to advocate and promote the awareness of citizen groups and to involve citizens themselves in a dialog. The Nederlands Fotomuseum is a leading player in this social process.

For example, it organized sessions with elderly people in their retirement homes to formulate new insights on migration history (an important subject in the Nederlands Fotomuseum collection), to entertain these people, and to learn their stories. The museum did this by asking them to share their photo albums, and sometimes acquiring them when appropriate, with their visual stories about migration. This way, the connection between society, city, and collection became intrinsic. Refusing to adopt an ivory tower position within its own field of competence, this museum expresses an ideological vision with strong ties to the local culture and with a dynamic, democratic, and practical vision of the relevance of its discipline. Mind you, this is not philanthropy or charity: this is about bringing your very "soul" into society and improving the social conditions around you. What if global corporations and commercial enterprises were to take the same path of commercial success through social engagement?

PART IV

Futures

Introduction

This final part focuses on what the future is likely to bring to the different systems of fine arts, design, and culture. Our explorations are conducted across different themes, different countries, and different methodologies, to embody the best spirit of multidisciplinary quest that this book aims to represent. The next three chapters highlight three crucial areas of development for tomorrow's evolutions of the field of analysis:

- the evolution ahead for design, including its interdependencies with fine arts

- the cultural production and consumption patterns in areas such as Asia and Latin America

- public opinion on the future of fine arts in leading countries such as the United States and the United Kingdom, as captured by means of online statistical rule-developed experimentation.

The focus of this fourth part moves across a number of very diverse areas, with analysis based on both qualitative and quantitative tools, aiming at finding viable answers to three key questions:

- What can today's "design for collectors" and tomorrow's socially oriented design teach us?

- What can we learn from the ways fine arts and creative industry systems emerge in new regions?

- What is the future of fine arts and "design for collectors" according to a qualified statistical sample of ordinary people?

In Chapter 7 we discover and use inspiring views and ideas by Bruno Munari and Roland Barthes. As recorded in their thought-provoking essays, design can be described as one extreme of a spectrum where:

- on one hand, fine arts represent a territory of free exploration with just aesthetic and reflective purposes (Munari)

- on the other hand, design itself represents an application of creativity within constraints and with utilitarian purposes (Munari), resulting in the production of sociological symbols with great impact on the collective imaginary (Barthes).

From this departure point, we discover, with the help of a case study, the recent notion of DesignArt, overlapping with the more general "design for collectors." This has been one of the driving forces of commerce in the field for at least the last five years. It is easy to demonstrate the recent strength of this specific facet of design in pure market terms by reminding ourselves of the Introduction to Part III. The TEFAF Design section, newly launched in 2009, made the news thanks to its first-day sale by the Philippe Denys Gallery in Brussels of a 1932 walnut and aluminum library case by the great Turin architect, Gino Levi Montalcini, to a French collector. Also important in terms of sales was a stunning 1983 fountain installation by Pol Bury on show on a Paris design gallery's stand. In this chapter we appreciate the value of classic design pieces by the likes of Italian architect Carlo Mollino.

In order to understand how design became part of the higher cultural scene, we go back to the 1930s, to the early history of the Museum of Modern Art, the first major institution to adopt design within its collection. We discover how the transition from mundane commodities to museum exhibits took place. It is vital to understand how, in recent years, the work of contemporary designers has been distributed, priced, and discussed as fine arts. I report on the design shows by Mark Newson at the trendsetting Gagosian galleries in London and in New York, and on the extravagant excesses of top talents like Marcel Wanders at the Salone del Mobile in Milan. We see how design lost its critical edge and how it is currently fighting to get it back. The focus of my critique is on Dutch design, and how in the last 15 years this went from healthy radicalism to commercial exploitation.

I love Dutch design and regard it as one of the most outstanding schools of applied arts in the world. Exceptional talents such as Paul Linse, who signed off the Schiphol Airport VIP lounges, and Piet Hein Eek, with the rigor and rough elegance of his scrapwood artifacts, are but two testimonials of its exceptional quality. Dutch design was however unprepared for the downturn. I sketch how initiatives are being taken in terms of urban presence (Red Light Fashion) and ideological leadership (Repair Manifesto) to get Dutch design back on its feet, and ready to jump into a next decade of success because of a more social orientation. From Dutch design I extrapolate conclusions and directions of a universal nature.

Chapter 8 moves to the future developments of nonwestern art markets. In order to look ahead, we investigate the present and the past. Within such analysis, I use the opportunity to gather a more profound understanding of the classic Eastern aesthetics and culture of the last century. We discover Japanese writers such as Tanizaki and Okakura, in an intellectual connection with contemporary art superstar Takashi Murakami, in the case study.

Beginning with an understanding of the cultural mindsets of wealthy Japanese collectors, we move through Chinese creative industry circles to land in the Indian high art market explosion and Brazilian street art. I strongly advocate the crucial value of long-term investments in local networks and creative communities by those business leaders and entrepreneurs who really aim at being successful in the long term in these countries. The advantage of a truly localized approach to new markets is demonstrated by the direct experience of Davide Quadrio, the Italian maverick who started from scratch a new way of doing business in fine arts, design, and consulting in Shanghai. Other players across Asia we meet include the Tokyo Gallery + BTAP (Beijing Tokyo Art Project), and Chemould Prescott Road of Mumbai. The thread throughout the chapter is to immerse ourselves in the history, the physical reality, and the social networks of the countries where we all aim to achieve business success, and to do so with the greatest authenticity and patience. Genuine commitment to local societies is utterly essential for business success in a global context that will increasingly dewesternize, recovering national and regional roots at aesthetic and cultural levels.

A quantitative exploration of the public perception of fine arts in the United States and the United Kingdom is the core of Chapter 9. The purpose of this chapter is twofold:

- It aims to offer cultural managers and museum directors an array of practical insights highlighting where public opinion in the United States and the United Kingdom sees the sector going.

- It offers positive quantitative validation to our qualitative knowledge base.

The research carried out resulted in a segmentation of respondents into more conservatively attuned and more socially engaged groups. In spite of the economic downturn and in parallel with the emergence of new geographies, US and UK audiences are still among the main recipients of what the fine arts system will generate in the next half decade. This final milestone will therefore close the circle of our geographic review, from the Far East to the advanced economies of the west.

This section has 13 "Findings:"

- Finding 26: Design for the future.

- Finding 27: Design for the cultural system.

- Finding 28: Let people co-design with you.

- Finding 29: Design the evolution of your design "grammar."

- Finding 30: Design to reconnect your vision to people.

- Finding 31: Reach people by truly understanding their culture.

- Finding 32: Reach local networks to be part of them.

- Finding 33: Reach local markets by creating ad hoc business models.

- Finding 34: Reach and nurture local creativity circles.

- Finding 35: Forget high and low: reach culture at all levels.

- Finding 36: General directions for the future of fine arts and "design for collectors."

- Finding 37: People's statistical preferences, explained.

- Finding 38: A detail is worth a thousand segments.

Along the way practical insights are provided to business leaders and brand marketers. First, the mechanics of design-driven cultural marketing is critically reviewed. This means that I particularly address the potential risks of adopting a strategy aimed at elevating design by showcasing it within the cultural system. Second, the relationships among vision, execution, and identity are explored in order to understand how design "grammars" can be managed over time by brand directors and strategic marketers. This is a crucial topic in the contemporary corporate debate about heritage, consistency, and policies. Moving to tomorrow's new local markets, there is no doubt that the future will increasingly require the ability to navigate different societies and diverse culture. The relevance of a deeper understanding of culture at both levels of historical studies and of field ethnography is highlighted at all levels and with the deepest belief. To conclude, the Rule Developing Experiment reconnects qualitative visions with the scientific measurement of statistics. This provides validation to the aggregated "findings" as well as, I hope, inspiration to those who will be tasked with defining the future of fine arts and design.

CHAPTER 7

Design: directions towards the future

Throughout the 2000s, "design for collectors," or DesignArt, described in this chapter's case study, traveled like a comet in the sky of contemporary art markets. Here, limited edition artifacts were traded for the wealthy as fine arts artifacts in terms of distribution and price. Beyond commercial success, this is just one of the facets of contemporary design. Even more important, in times of economic constraints, is the notion of urban design, where public art offers the opportunity to rethink infrastructure into designed city landscapes. In fact the worlds of fine arts and design seem closer than ever today in the necessary drive to redefine the cultural climate by turning their talents to social themes. After all, a number of important past design movements and ideologists have worked to build up the profile of this discipline as part of a larger aesthetic universe where fine arts methodologies and humbler objects converge in the beautification of the everyday.

In this chapter, we review various trends and explore how design is changing, with the ambition of once again provoking change in the future.

The business question behind this chapter

What can today's design for collectors and tomorrow's socially oriented design teach us?

An introduction to the findings in this chapter

What are the mutual relationships between fine arts and industrial design? Seeking answers, we adopt a (perhaps scholarly incorrect but) practical scheme. This is the simple picture painted in the entertaining

1971 book *Artista e Designer* by Italian artist, designer, critic, and creative thinker Bruno Munari. Munari, also associated with Futurism and with ISISUF, was one of Italy's most highly regarded innovators on these topics, and his cultural legacy is still strong. In his book, he makes our life easy by sketching a very schematic classification. In the past, according to Munari, things were simpler:

- works of fine art were created for exceptional purposes of an aesthetic and ideological nature

- the design of everyday things was just part of the industrial production, mostly for the masses.

Within this simplification, fine arts is the realm of self-expression, where the classic rules of Renaissance painting and thinking still hold true, even when denied by avant-garde artists. While the artist is as much as possible concerned with self-expression, the industrial designer is involved in multidisciplinary team work to achieve practical goals, translated into the mass production of "everyday multiples."

Further to these two concepts, Munari introduces the notion of "stylist." This is an intermediate professional who applies some fine arts principles (for example, the potential pursuit of decoration and embellishment beyond function) to the functionality of applied arts. Munari's appraisal of "stylists" was extremely critical, almost to the point of sarcasm. This third element in Munari's scheme leads us to:

- the "artist," who is led by experimentation and beauty (or the deliberate negation thereof) in the realm of aesthetics and fine arts

- the "designer," who is led by rational teamwork and multidisciplinary insights, in the realm of applied arts for industrial production

- the "stylist," who applies form to function in a fashion-driven, decorative manner, in the realm of "trends."

This abstract "working sketch" by Munari is a perfect way to establish a foundation for this chapter on design. Munari's work was published in the same year as Germano Celant's retrospective on Arte Povera and its counterpart movements in the United States. The current status of relationships between fine arts and design is much more complex and rich than in 1971, and this chapter brings us to the challenges of today's design. Here's an example from 2009. Larry Gagosian is one of the leading art market leaders in the world. The trendsetting success of his US galleries on Madison Avenue and on 24th Street in New York matches the commercial success of his UK galleries in Britannia Street and Davies Street, London. His latest opening in Rome provided him with an opportunity to enter the Italian collector market as well. His selection of contemporary art shows is as sophisticated as it is varied: At the time of writing, it encompassed an anthological show of marble sculpture and a retrospective by Piero Manzoni in New York, shows by American legend Cy Twombly and by Japanese trendsetter Takashi Murakami in London, and the launch of a new fragrance, in reality a new conceptual art project, by Italian rising star Francesco Vezzoli in Rome.

More important than the location of his galleries, it is Gagosian's power, which is close or equal to Charles Saatchi, which makes him a reference point in the art world. In the contemporary art sector, what Gagosian does is an indication of where the market is going. I selected Gagosian's commercial promotion of designer Mark Newson as the most representative issue to introduce the notion of DesignArt (see the case study). Following a show at Gagosian in New York, Newson, a star of the design world, had a London show that comprised a collection of unique pieces in marble, a sculpture set of top designs available for art collectors as well as design lovers via Gagosian's gallery. The press release indicated the timing and the dates of opening and closure of the show, including the opening reception for the artist, as usual for all art gallery shows.

Wait a minute. Usual? Mark Newson, an Australian born in 1963, designs watches, clothes, jewels, interiors, and objects, from home products to airplanes. He is very popular among industrial designers for his round shapes and for a style that was defined as "biomorphic" by critics of design aesthetics. He has worked for design household names such as Cappellini, Ideal Standard, Alessi, Tefal, and Magis, as well as Ford and Dom Perignon. In essence, he creates high-end series, limited editions and prototypes for the best brands in the world of design and luxury. This makes Newson a very talented industrial designer with a strong lifestyle orientation and a successful track record in his profession as a commercial designer of projects for business enterprises. Newson's Carrara and Bardiglio marble shelves and his Striato Olimpico marble table come – at least theoretically – from another world: that of applied arts. To recall Munari's classification: what would a "designer" like Mark Newson do, in a gallery that usually displays "artists' work?"

This chapter dissects the "why" and the "how" behind the meeting between an industrial designer such as Newson and an art gallerist such as Gagosian. Mind you, we do not address design in terms of its mass-production-related nature, as Munari did. In this review, we focus on the way designers and design interact with fine arts systems, to find out how design is turning back to society again (as it did in the past). It must be clearly understood that we share Munari's viewpoint: The risk for designers entering the realm of fine arts is to fall back into just "styling." In this respect, DesignArt could be described as a reactionary bubble of overvalued design distributed through the gallery system. Our "findings" map both the status quo and future lines of development in these fields, as well as providing inspiration for business leaders and brand marketers:

- Rethink your design manifestations as potential future cultural icons, to create the conditions for brand value creation over time.

- Search for crossovers among diverse cultural domains, for example, everyday consumption as opposed to cultural circuits, to connect and cross-pollinate in order to identify different ways to promote and profile your brand.

- Always leave room for the end user's contributions in the co-creation of your design manifestations.

- Find your own "vernacular" in terms of signature aesthetics, and develop it over time with consistency and coherence.

- Move the focus of your design practices from products to societies, to be part of urban cultures, anticipating the next trend developments.

It is time to explore the work of designers and urban artists. We present our five "findings" and their implications for business leaders and brand marketers.

Finding 26: Design for the future

Create design manifestations that will last for years

History is the driver behind the fascination of specific collectors for what is remote in time, and blessed by the unique value of the aristocracy of all things past. As the world is on the edge of financial excess and an uncertain future, revisiting our past and re-examining our roots is part of an inevitable process of rebirth, because it is in our past that we will find the legacies and the bases to lead us into the future. In this environment, the first impulse of the world of fine arts might be to stop: stop for a while the frantic rat race and the gold rush of the contemporary (a marketing process that is commercially dominating entire nations and continents, from China to the Arabic countries, from DesignArt to design fairs). It is in our past Old Masters and classic design masterpieces that the best market value will be found for collectors who appreciate solid and reliable investments. This "finding" develops the design side of this theme.

Our journey in the world of classic design for collectors begins with an important record price set at an art auction. In a 2006 interview with *Art + Auction* Ms C. Grajales defined herself as a "decorative arts consultant, dealer, and design curator." Her fame is due less to something she did, however valuable her influence in the design world might be, than to something she bought. In June 2005, she purchased a 1949 table by Italian eclectic architect Carlo Mollino at Christie's for US$3.8 million. This was the highest price paid for a design piece at auction, and demonstrated how art and design are converging in commercial terms. The value of the custom furniture pieces designed by Carlo Mollino continued to increase before the economic downturn. The story of Mollino's talent is told in a book by Fulvio Ferrari and Napoleone Ferrari, *The Furniture of Carlo Mollino* (2006). It reproduces an invoice by the Apelli F. – Varesio L. dated 20 January 1954, to a Mr Colonna. The latter, a lucky Turin-based gentleman, had ordered one table, six chairs, one hanger and one functional part for a kitchen at the luxury price, at least for the time, of 400,000 Italian lira, the equivalent of more than ten times the average monthly salary. So we have a completely theoretical, yet highly remarkable example of value – increasing, in some 50 years, from 400,000 Italian lira (perhaps equivalent to US$100,000 currently) to potentially up to US$3.8 million. Is this the commercial power of the designer as an author? Is this still design, or is it something else? After all, the art museum GAM in Turin held a full Mollino retrospective, evidence of the artistic value of his signature. Is this trend going to continue?

As well as the exceptional case of Mollino's pieces, there is no doubt that "historical design," for the sake of a better definition, is strongly *en vogue* at the time of writing this book. In the worst weeks of the 2008 economic downturn, the press office of TEFAF announced the launch of TEFAF Design 2009. This new, dedicated section of the fair was created to focus on offerings by world-class players in the relatively restricted domain of design galleries. These included Sebastian + Barquet (New York and London), Galerie Downtown-Francois Laffanour, L'Arc en Seine and Galerie Eric Philippe (Paris), Philippe Denys (Brussels), and Galerie Ulrich Fiedler (Berlin).

One of the real stars of TEFAF Design 2009 was the selection of historical furniture by Dutch architect H. P. Berlage, as part of the permanent collection of the *Jachthuis* (or hunting lodge) built by the Kroeller-Mueller family in the late 1920s. The interest of TEFAF design collectors is in established names such as Charlotte Perriand, Alvar Aalto, and Peter Behrens, and in design pieces dating from the early 1900s to the early 1980s. One of the star signature star pieces was a "Conoid Bench" by George Nakashima for the Rockefeller Japanese House in New York, a handmade masterpiece from the mid-1970s. Just like Mollino's table, these pieces were not designed to be mass produced or mass consumed. It is clear that both designer and commissioner invested important amounts of time and money in most of these works. On the other hand, the fact that these artifacts were selected to share the spotlight of Europe's most sophisticated fine arts fair, one that began by focusing on ancient masterworks and antiquaries, is the best indicator of the true, solid value of classic design in the fine arts systems. The convergence of these two worlds, product design and fine arts, is the very start, at a historical level, of the contemporary DesignArt phenomenon.

From the perspective of business leaders and marketers, the benefits of a strategy focused on long-term design value are self-evident from a viewpoint of brand equity and reputation over time. The main challenge raised by the analysis of classic design pieces and their market status is: Are your designs created for longer-term brand value generation? If the answer is "yes" then your brand is on its way to establishing itself as a design leader over the decades. If the answer is "no" it might be worth thinking again about this purely theoretical, yet surely appealing, potential opportunity area.

Finding 27: Design for the cultural system

Find the right crossovers to profile and position your designs as cultural icons

Raising the perceived value of brand manifestations from normal everyday consumption to the realm of higher culture is an important long-term asset for brands. For a designer, in the last 20 years especially, being featured in a museum became an important moment of validation in professional career terms. For a brand, it is a golden opportunity in terms of reputation, profiling, and PR future assets. A product in its mundane environment is just a commodity. The same product under the spotlights of a major museum show will be charged with the charisma of culture and – possibly, in time – with the magic patina of history. This second "finding" explores how design can reach museum status and how such status can positively work or fail to deliver success. We initially return to the 1930s when this took place for the first time, then move to today. Our story begins in New York City. Isn't the Big Apple home to the best contemporary museum to use as a stage for your brand design manifestations?

Perhaps no other institution since the beginning of the modern age has been as influential as the Museum of Modern Art (MoMA) in New York City. Here the curator of design, Paola Antonelli, oversees the policy of one of the cathedrals of contemporary applied arts. Her selections come close to making or breaking careers in terms of prestige and credibility. The roots of this powerful status lie in the very history of MoMA: this is the museum that brought product design into higher culture. How did such a revolution happen? It began with the multidisciplinary drive of its founding fathers. In the 1930s, the intellectual work of Alfred H. Barr, Jr. and of his associates, Philip Johnson and Henry Russell Hitchcock, Jr., was crucial in establishing the "international style" in architecture. This was expressed as a new way to look at the challenges of design and decoration of buildings. The first manifesto of these sophisticated scholars was the book *The International Style* (by Hitchcock and Johnson), published in 1932. The main event that made it happen however was the "Modern Architecture" show at the MoMA in 1934. This exhibition established international style as *the* style of the times. The actual foundation of a new vision of architecture, from skyscrapers to villas, was launched in the severe rooms of MoMA. Introducing design at MoMA was the natural next step for Barr and his colleagues. As an ideal multidisciplinary follow-up to "Modern Architecture," three major design shows took place: the second, "Machine Art" in 1934, was the first time that everyday product designs were displayed as "cultural artifacts" in a museum of fine arts. This is a status to which any designer or brand should naturally aspire.

Today, the burning question for brand marketers is: Is there an optimal process to enable your design artifacts to be selected, exhibited, and profiled by museum curators? The answer is, unfortunately, "no." At least, not in the sense of ISO standard templates. Of course, it is possible to describe what happens in a generic step-by-step path from product concept to a museum show. The process is pretty straightforward and simple:

- Strategize from the beginning your design policies to incorporate the objective of possible adoption by cultural institutions.

- Select specific star product lines or create ad hoc research projects, to be produced in order to support such cultural ambitions.

- Do your homework in terms of studying in depth the policies, the history, and the marketing of the museum or cultural institution you want to show your design.

- Carry out the design to the current quality standards that will appeal to cultural curators and design critics that have an affinity with your trade.

- Ensure that you are in a position to be featured on the cultural circuit as a dignified player, with reputational benefits in your own category or industry.

Simple, isn't it? Unfortunately – or fortunately, actually – things are not quite that simple, because cultural profiling does not necessarily always deliver the expected outcome. Instead, it takes a lot of management talent in terms of individual taste, lobbying, networking skills, and similar personal assets, to pursue cultural visibility for a brand through design. A purely mechanical approach will simply result in failure, at best with a loss of investment, at worst with reputation damage.

To see how things can go wrong, we move from New York to the Netherlands, and a recent Dutch fashion phenomenon. Since the early 1990s, the Netherlands has adopted a state-supported strategy to make Dutch fashion "happen." Dutch academies created a number of potential designer brands, mostly in the name of the actual graduates. The work of young Dutch designers was supported by sponsored shows in Paris and Rome, and by an intelligent system of combined publishing and cultural marketing efforts. The museum system was vital as it absorbed and culturally elevated the creations of young graduates straight from the academies, with what appeared to be a rapidly growing enthusiasm. Everything was seemingly in place to grow Dutch fashion as a sort of collective brand to success through museums and art systems. Then, simply put, nothing really happened. Vitally, what was missing was an overall fashion system with back-end logistics, technical marketers, and strategic brand design expertise. Also missing was a genuine tradition of everyday fashion in society. These fundamental gaps contributed to the substantial failure of the initiative in terms of global relevance in spite of the excellent cultural strategies deployed. To date, the only real Dutch fashion brand of the 1990s (beyond the early master Alexander van Slobbe) is Viktor & Rolf, the conceptualist maverick duo who played the cultural museum game at its best but without getting really involved in the larger subsidized promotional system.

This example confirms that – no matter the magnitude of resources – no standard process exists for secure business success in moving design into the cultural and museum fields. One thing can be stated, however: Design that goes into museums should always be validated by real customer demand in the real world and not be just the outcome of "arty projects." This is because it is consumption that dignifies design, since its nature is functional at heart. Finally, the proof of the pudding for design always lies with people like us, who adopt it and use it every day, or do not. Here, authenticity and a reality check rule.

Finding 28: Let people co-design with you

Open your brand to the power of bottom up participation

In 1908, Austrian architect Adolf Loos published his essay *Ornament and Crime* as an accusation against the art nouveau style of those years, and its erotic decadence. The *Gesamtkunstwerk* (universal artwork) principle criticized by Loos entailed rigidity and the absence of any space for future growth or personalization. According to Loos, such ultimate completeness might be unbearable to the rest of us. In 2002, Hal Foster, the American cultural and art critic, extrapolated Loos's conclusions to the present-day design industry. In his essay, *Design and Crime*, Foster highlighted how the negative critique written by Loos is perfectly applicable to today's design standards. Contemporary designers engage in the creation of "signs that carry goods" (and not, notice, the other way around). Designers ultimately engineer total lifestyle aesthetics. This is about the same drive to rigid completeness emphasized by Loos in his reflections. Mind you, this is not about Depero's ideal reconstruction of the universe: That was about vision. This is about pervasive logos and branding markups. What are the best examples of such "crime"?

An ambition of controlling each stylistic detail – from fashion to accessories, from department stores to branded retail – pervades luxury brands. This is a zero sum game: on a long enough timeline, there is inevitably no space for growth in the strategy to impose signs, symbols, and icons on the masses. In Chapter 4 we saw how the experience of Futurism could be translated into the opportunity to rethink the universe according to your own vision. We also discussed how this could apply to contemporary lifestyle brands, and said how it would be mistaken to translate such a "universal vision" into "oppressive aesthetics." The main focus of attention for business leaders and marketers is to understand the difference between the formulation of an ideological platform, and its actual design implementation:

- At the level of "vision," brands should indeed rethink their universe in a more encompassing fashion.

- At the same time, such a deepening of the ideological dimension of brands should not be matched by a tightening of aesthetic control in their design.

This "finding" has a great affinity indeed with the challenges facing contemporary luxury brands. We therefore focus on premium categories as *the* field of reference for our analysis of its contemporary business implications. Once again, however, we start from history, from a masterpiece of 1920s architecture, designed shortly after the publication of Loos's essay. It is a story of superior design, and to some extent, of surprising user discomfort.

As mentioned earlier, a centerpiece of TEFAF Design 2009 was the 1920s/1930s *Jachthuis, Saint-Hubert*. It was designed and erected by Hendrik Petrus Berlage in what is now the Hoge Veluwe national park. The project was carried out for the art patrons and wealthy commissioners, Anton and Hélène Kroeller-Mueller. The *Jachthuis* was a hunting lodge to complement the couple's sophisticated lifestyle. It was designed to be consistent in all its smallest details, from the patterns on door handles and textile prints, to the plan of the building. The perfection of this *Gesamtkunstwerk* went to the extent of having exactly the same time on each clock in the house. The drive for completeness was total. This building surely stands as a climax in excellence in the fields of European twentieth-century architecture, design, and fine arts.

Given these premises, you would think it would be a great pleasure to live in such a house. During the 1930s economic downturn, Kroeller-Mueller and his wife had to spend substantial time in the *Jachthuis*, since the financial crisis affected their sources of income, which were predominantly connected to international trade. Historical documents surprisingly signal a great degree of discomfort, especially suffered by Hélène. This is a first sign that the perfection of Berlage's design might not have translated into an optimal user experience. Also, we have the informal comments by civil servants who currently inhabit the *Jachthuis*: they are less enthusiastic than you might think. In spite of the paradise of its surroundings and the sophistication of its design, the *Jachthuis* appears to be just too – complete. The impression is that people who have the rare opportunity to really live in this building feel suffocated by the rigidity of this environment, made and conserved strictly according to its design specifications.

Of course, when a genius like Berlage realizes his own vision, the outcome is a national monument of rare beauty. Regardless of whether people want to live in it or not, it is clearly worth preserving and studying. On the other hand, moving on to "average" luxury designers or lifestyle brands, things are likely to change. For example: isn't luxury retail boring in the end, being all the same the world over – brand after brand – in terms of format, color schemes, and design? Isn't the drive towards complete control exercised by luxury brands ultimately unappealing to free spirits and thinkers, in spite of advertising campaigns saying the opposite? The alternatives to the design crimes perpetrated by luxury brands, and by brands in general when adopting conventional marketing and design techniques, are simple:

- Stop thinking in terms of dictatorship of aesthetics, and first redesign your own vision of the world at an ideological level, according to a more democratic relationship between people and your brand.

- Create a more flexible, open, and interactive brand universe where people – customers, stakeholders, communication audiences – have room to co-design, co-create, and truly participate in the expression of the brand in its material forms.

Based on these conclusions, we arrive at a challenging crossroads for business leaders and brand marketers, especially in the luxury and premium lifestyle categories. The questions for them are, Does your design strategy leave space for people to co-design its forms while demonstrating a comprehensive vision of the world? and, Does your design strategy pursue standardized formats, void of any personalization or localization?

Given the proliferation of luxury branded environments, from department stores to hotels, this second question is a crucial challenge for business leaders and brand marketers, one that should be addressed in depth, by designers and by their commissioners, the brand custodians of the aesthetic "grammar" of such luxury premium brands.

Finding 29: Design the evolution of your design "grammar"

Find your own vernacular, nurture it, and evolve it in time

This "finding" focuses on the necessity for business leaders, top marketers, and design directors to manage over time the cultural DNA and the "design grammars" that govern their brand. The underlying theme is the need always to maintain coherence and consistency between the design for a brand and the vision that such design should manifest. Starting from a more general and generic context, three questions arise:

- How far is your design strategy still connected to the original vision that generated it in the first place?

- Are you managing your design grammar with coherence and consistency over time?

- Are you well connected to people's aspirations and values, and ready to respond to changes in the field of society and culture, by design?

The recent history of Dutch design, from its blossoming to its ideological crisis, is the main focus for this "finding," with the design collective brand Droog Design, and its evolution as our main reference. Droog is vital to Dutch design: it is one of its flagships. Although Droog's history is relatively recent, it is important to international design as we know it today. In essence, we will learn not from best practice but in the case of Droog Design, from what went wrong, and how it went wrong.

In the mid-2000s, no one and nothing could stop the flamboyant growth of Dutch design. The aesthetic experimentalism of this talented nation was publicly praised by the likes of Paola Antonelli, MoMA curator, and Italian design legend Alessandro Mendini. At the 2008 Milan "Salone del Mobile," the defining moment of the international design agenda, the avant-garde Zona Tortona area, was overwhelmed with installations, concepts, and products, all made in the Netherlands. Superficially, the feedback was totally positive. However, informal comments and person-to-person discussions in the creative industry made it clear that spring 2008 was already too late for Dutch design. What was going wrong? It was evident that the richness and authenticity of this design vernacular had been progressively replaced by marketing gimmicks, vague concepts, and an overall sense of arrogance.

The contemporary trendiness of Dutch product design at international level had one of its most important originating points in 1993, with a presentation in Milan by the Droog Design collective. Initiated by Gijs Bakker and Renny Ramakers, Droog was introduced in a collective show at Paradiso, the alternative rock culture night temple of Amsterdam. Back then, at its origin, Droog's vision of the world was described as:

- raw

- rough

- humorous (with dry humor)

- conceptual, with a clever twist in terms of product use or of the materials used

- to a large extent, antagonistic to contemporary consumption culture and mass marketing.

That was the grammar of Droog's "vernacular aesthetics" in the early 1990s, when it rose to fame. This is an ideological footprint that carefully matched the nature and character of Dutch society and culture of those years. The culture of Netherlands has historically been a unique mix of tolerance and free thinking. Not known for its aesthetic finesse, the Dutch vision of the world is based on Calvinistic transparence, individual leadership, and energetic drive to individual enterprise. The original power of Dutch design was in its Dutch vernacular grammar: a mix of humor, carefully planned imperfection and antidecorative edge, all glued together by the natural exuberance of a generation of designers by nature optimistic and entrepreneurial.

Through the years the Droog collection expanded, and so did its distribution, from Design Republic in Shanghai to its own shops in Amsterdam and New York. In a relatively short time, Droog became one of the most successful icons of DesignArt. Then, the economic downturn came, and the perception of Dutch design rapidly changed. As a notable example, Michael Cannell's article "Design loves a depression" in the *International Herald Tribune*, online on 3 January 2009, used a number of Dutch design names and examples to attack the frivolity of design as we knew it before the credit crunch: Rem Koolhaas and his 1995 book *S, M, L, XL*; Hella Jongerius and her US$10,515 "Polder" sofa; and especially Marcel Wanders, one of the brightest talents to emerge from the lowlands, with his 2005 party at the Milan fair. There, Wanders displayed his girlfriend – the choreographer and dancer Nanine Linning – suspended upside down from a chandelier, serving cocktails and drinks to the guests below.

Cannell's article was severe, and overdue. On their way to success, both Dutch design and Droog gained in world popularity, business success, and great media exposure, but they also lost a number of crucial aspects. They lost their cutting edge for example, leaving the cultural design scene of the country void of a critical voice. They lost their position of underdogs, to become highly integrated in the international design establishment. They fell into a neutral zone of comfort, pampered by national academies and museums, and the Dutch media. What had been a genuine spirit of antagonist rebellion turned into a negative mannerism that could be interpreted as cynical commercialism. Dutch design embraced the consumer society with natural passion and with unique talent: this marriage, however, stifled creative talent in a move towards generic strategies. As seen in Milan 2008, the (commercial) success of younger generations was almost desperately pursued by the entire system as a token of self-recognition. Not a comfortable position to be in, at the edge of the economic downturn.

The question then is: How can a genuine vernacular grammar for Dutch design be reclaimed? Currently, the options are open for Droog and for Dutch design in general, either to strategize their role within the design processes of new geographies, from Russia to China, and act as consultants (genuinely contributing to progress in new countries), or to reinvent their own business model. One thing is certain: Dutch design, the international star of the 1990s and 2000s, will have to undergo a process of soul-searching, for its renaissance and rebirth. All the enabling conditions, beginning with the natural optimism of the professionals involved, are there to make the 2010s a successful decade for Dutch design – with the condition that Dutch designers be capable of connecting with people's newly developed socio-cultural values. In our next "finding," we will see examples of how this can be done.

Finding 30: Design to reconnect your vision to people

Putting society and the city back at center stage

The history of Droog Design highlighted the limits of DesignArt as an approach to design. After its years of complacency with commercialism, reestablishing the credibility of design is one of the main challenges in the next decade. Having focused mostly on the styling excesses that served capitalist bankers and MBA-educated managers eager to spend their bonuses in the acquisition of lifestyle markers, design will have to move beyond DesignArt and recapture the ethics that made it socially relevant in the dreams of the founders of MoMA, in the Great Depression programs of urban and public investments, and in the heart of contemporary culture.

Earlier, we mentioned the power of design and art within cities. In this "finding," we raise two questions of future opportunity for the benefit of civic servants and business leaders:

- How can design enable change processes in urban areas, acting as an engine of regeneration and rebirth?

- How can design be the engine of new ideologies, generating visions with much-needed political impact?

These two questions represent fundamental junctions and areas where the future of design will unfold. Are there early signs of the change ahead? An example of how design helped to initiate change in the urban area is the Red Light Fashion Amsterdam project. Here, the Amsterdam municipality worked to achieve urban recuperation and civic regeneration of urban areas. Mariette Hoitink, founder of design consulting firm HTNK, created a program that has been running for a year, and will do so for a second year, thanks to the support of the public commissioner and multinational sponsors. Red Light Fashion Amsterdam involves 16 Dutch fashion labels, including Jan Taminiau, Bas Kosters, and Mada van Gaans, plus CODE (a fashion design gallery), and Petrovsky & Ramone (fashion photographers). The group of young fashion makers and lifestyle trendsetters was given access to a number of rooms in Amsterdam's red light district, converting selected world-famed red light windows into fashion design displays and workshops. Living and working everyday in the red light district, these designers gained media attention and some degree of promotional fame. At the same time, their designs were part of a larger project to improve urban well-being and quality of life in a controversial area of town.

Red Light Fashion Amsterdam is an admirable initiative. However it has a limit: its social connection to the city does not go beyond marketing communication. This is because its design scope and content does not address the real ideological issues by posing questions or making statements about the – somehow, and to some extent – reactionary closure of part of the city that for hundreds of years has been a free lifestyle area.

Where then can we find more radical manifestations of new design thinking? To do this, we move to the south of the city. Here, in a round chapel, is the head office and exhibition space of Platform 21, a creative laboratory founded by Premsela and ING Real Estate. At the time of writing, Platform 21 has just launched a "Repair Manifesto," with much fanfare in the blogging communities and design media, and with an important forum in Milan, during the 2009 furniture fair. Structured in 11 statements (just like the manifesto of Futurism and of course, with all due humility, our own *Golden Crossroads* manifesto), the Repair Manifesto offers a complete new vision of the central role of users and owners of artifacts in a whole new future culture of consumption. The 11 statements in the manifesto are:

1   Make your products live longer!

2   Things should be designed so that they can be repaired.

3   Repair is not replacement.

4   What doesn't kill something makes it stronger.

5   Repairing is a creative challenge.

6   Repair survives fashion.

7   To repair is to discover.

8   Repair – even in good times!

9   Repaired things are unique.

10 Repairing is about independence.

11 You can repair anything, even a plastic bag.

Here, Dutch design is challenging itself and the world to rethink its operations and presence within society, from a cultural viewpoint. More politically relevant than Red Light Fashion Amsterdam, this simple document is a perfect ideological springboard for the next phase of Dutch design. This is the best of the spirit of Dutch culture: against the mass culture of consumerism, vibrant with a spirit of practicality so extreme that it reaches almost a lyric vein of poetry, and with the dry and raw edge of a single-minded opinion. The penetration of this document to international design thought-leading debate and public opinion professional circles has been fast and powerful, with a huge amount of blogs reproducing it.

The challenge for all of us, including business leaders and corporate management, is that there is a lot of potential change coming from this design viewpoint right now. Questions arising from this journey so far include:

- Do your enterprises and corporations creatively rethink the way your design excellence is brought into cities, cultures, and societies, to become an agent and ultimately a protagonist of positive change?

- As well as the everyday styling tasks assigned to the various design teams in their own areas, are your enterprises and corporations using their own design strategic visionaries to rethink what design is and how design works?

- Are your corporate design teams working to enable the necessary cultural change that will position your brands and products at the head of emerging social trends?

These questions are among the most crucial challenges that business leaders will face in the next couple of years. This field might look secondary when compared with financial or corporate governance challenges. However, design and its cultural and social relevance is one area where enterprises will find opportunities to transform the downturn into an opportunity for renewal and renaissance. Missing this chance would not mean to just return to the world of hyper-styled, price-inflated objects we have known in the last decade – those times are not coming back for a long while. Missing this chance to rethink design and to think through new design could mean ending up in the bonfire of vanities where the brands and names of an age come to their natural end and eventually disappear. The chance is there, for every single enterprise and corporation, to make a difference. The risk is also there to ultimately fail in maintaining any social and cultural relevance, and most likely vanish.

Wrapping up: from DesignArt to the future of design

From the 1970s classification by Bruno Munari, design has evolved dramatically. It was the supposed torchbearer of applied creativity to problem solving with an implied social focus. The rise of design objects as museum artifacts can be read as the elevation of designers to the role of fine art creators, as the natural step of applied arts coming closer to the cultural relevance of fine arts. It is a long path that was validated, in the 1930s, by the ground-breaking MoMA shows. Why has design been so relevant in the last decades within cultural systems, to the point of generating a DesignArt bubble that took center stage? In search of answers, we go back to the 1957 essay in *Mythologies* by Roland Barthes, the French philosopher, on the Citroen DS and its intrinsic design value(s). In this classic book of popularized semiotic analysis, we can extrapolate a few key conclusions that are still valid today:

- *Objects are magic manifestations of the cultural climate that generates them*: This is where the relevance of design as a creative force of society emerges in terms of its symbolic impact.

- *The choice of materials, finishing, and design solutions in successful objects relates to the symbolic level of cultural processes*: From this viewpoint, designers are the makers of society not just at the material level but at a higher degree of symbolic power, one that transcends the professional limits of their mundane everyday activities.

- *A car in 1957, and an iPod from the first decade of the 2000s, are respectively the modern and contemporary equivalents of a cathedral in the Gothic age*: Such artifacts are the symbolic manifestation of a culture and its spirit, as conceived, designed, and assembled by the mostly anonymous contribution of dozens of creative talents.

We can deduce from these considerations that designers are vital to the transformation of visions and ideas into offerings of a tactile and tangible nature. Within this environment, it is a natural conclusion to see the contemporary designer being elevated to a new status: that of nonspiritual, nonreligious *shaman* of a society of mass consumption. However, we should note how, from a more commercial and marketing viewpoint within the environment of fine arts systems, the rise of DesignArt represented the simple opening of new territories of exploitation and profit generation. Such opportunities exist at the intermediate level between mass production and the uniqueness of fine arts. A new market was created with a more digestible offering than overtly intellectual and semantically layered artworks (think of Arte Povera or conceptual art). This is as far as the success of DesignArt goes.

The focus of this chapter was however not the further exploitation of these phenomena and processes, but the next steps to move beyond them, and to understand how "design" can improve its functional role in the new societal organizational models and cultural mindsets that the economic downturn will determine. Things will never be the same: not for us, not for design. If design can move beyond the stylistic climax of DesignArt to take a new strategic role, paralleling the quest for social meaning and critical thinking of fine arts, then important opportunities will be created, and new chances will be within reach. What can business leaders and brand marketers do to take advantage of this dramatic change for design? These are the proposals derived from our five "findings" in this chapter:

- *Long-term thinking as opposed to quarterly profits and bonuses is needed more than ever*: Rethink your brands as potential future collectibles, to ensure value is delivered over a long period. This "finding" reminds business leaders and brand managers that the existence of their ideas and offerings is not limited to the next quarterly review by analysts: they might extend and widen its value for another 25 years.

- *It is wise to look for unexpected crossovers between culture and consumption, and vice versa:* This second "finding" stimulates ways to use culture to elevate everyday commodities to the status of museum artifacts – with a strong warning about the need to prevent bastardization of cultural elevation with short-term marketing tactics.

- *Leave space for the end user in the co-making of your designs*: This specific finding counterbalanced the valid indication in Chapter 4 of the need for more universal ideological visions by brands and enterprises.

- *Find your own vernacular in terms of signature aesthetics, and develop it with great care over time*: Here we urge business leaders and design directors to extract its own proprietary grammar of design, and to nurture not only its immediate commercial exploitation but also – more than anything else – its evolution over time.

- *Move your practices from products to societies, and be part of urban cultures*: We addressed the potential future of design beyond DesignArt, offering two examples from the Netherlands to policy makers, urban leaders, and cultural agents within society.

The bottom line of this chapter is that in the last five years DesignArt has been *the* emerging force in the fine arts. "Design beyond DesignArt" will pick up the need for change in societies and cultures, and act upon it quickly and effectively. This will constitute a major opportunity for business leaders, brand managers, and public policy makers as well as offering a unique chance for designers to stay relevant and to further strengthen their position as intellectual, media, and creative leaders in our future society.

DesignArt: What was it all about?

In existence since the late 1990s and reaching its peak in the mid-2000s, DesignArt is a manifestation of pre-economic downturn luxury. In these terms, it created a new segmented offer for aspiring collectors. DesignArt was also considered to be proof of how design had finally reached a commercial and cultural status comparable with fine arts. The art world took note, and welcomed design in its collections, galleries, and museums.

In DesignArt, exquisite artifacts in very limited editions are created by top names from the design and architecture world. Distribution takes place through galleries, either fine arts galleries or dedicated design galleries. Established fine arts galleries showcased design classics, placed alongside paintings, installations, and sculptures by leading artists. Ultimately, DesignArt can be described as design as luxury, as it used to be before the credit crunch and the Lehman Brothers collapse. This indisputable design trendiness didn't just begin by accident. For quite some time, much design was carefully positioned, staged, and painstakingly analyzed by writers in the know. At its peak, in spring 2008, the notion of DesignArt was discussed in a number of ground-breaking articles in the *Financial Times, Corriere della Sera,* and other newspapers, making it a major subject of debate in the main trendsetting media. Debate helps to shed light, dispel irrelevance, and make fascinating what might otherwise remain just interesting. DesignArt was on the Internet, and was therefore propagated by virality, buzz, and the new media. New websites such as www.20ltd.com offered limited editions and exclusive items not available anywhere else in the world, by contemporary masters like Zaha Hadid and Marcel Wanders. In the late 2000s, DesignArt was discussed as the new form of collectionism, as the next engine of luxury artifacts and as the hyper-consumerist realm of charismatic masterpieces. Then, the economic downturn came, a new world began to emerge from the ashes of the financial craze of the 2000s bubble, and design had to look to its roots once again.

CHAPTER 8

Regional gods: the emergence of new geographies in the world of fine arts and creativity

It is a given that today we live in a global market and – however damaged – in a global economy. The corporate drive to globalization has created greatly uniform offerings, with good benefits for consumers worldwide. People, however, are not just global marketing targets. The fine arts and applied arts sectors, with their nuanced cultural distinctions, will be our opportunity to find again the local dimension of emerging art markets. Here we introduce a number of insights into the new countries of fine arts and design: Japan, China, India, and Brazil. Parallel with this, the chapter moves from the luxury of Japanese high culture and wealthy individuals to the vibrancy of street art in underprivileged Brazil. Across our five "findings" consideration moves from the understanding of the local history of culture in Japan to the challenges ahead in the Chinese creative industry scene, including and beyond the world of fine art itself. What more inspiring way is there to continue our path of search and discovery across the world of fine arts than by looking at different regions from the viewpoint of creativity and its processes?

The business question behind this chapter

What can we learn from the ways fine arts and creative industry systems emerge in new regions?

An introduction to the findings in this chapter

So far, we have discussed various facets of the world of fine arts: its protagonists, its movements, its communities, and its museums, its place in high art and applied arts or design. Concurrently, we have presented cases and stories from the western modern tradition and contemporary markets. This chapter introduces new geographical areas for global fine arts – Asia and Latin America – and advocates the need to study in depth the texture of cultural differences. The relevance of the "rest of the world" is increasing from all viewpoints, not least the commercial side of contemporary art and design. As a testimony to the relevance of such growth, we can refer to a source of great authority: every year TEFAF releases an important report on the art market. For the 2009 edition, Dr Clare McAndrew, founder of Arts Economics, directed a research project on behalf of TEFAF, exploring the new global dynamics of fine arts. The results were encouraging, as summarized in seven points below:

- The art market in 2009 is a global marketplace, with new collectors in emerging economies – China, Russia, India, and the Middle East – which might help redress the impact of the downturn in advanced economies.

- This new wave of collectors is divided into more traditional and more status-oriented segments, with the latter purchasing in the contemporary art sector for reasons of investment, prestige, fashion, and national pride.

- At the time of preparing the TEFAF report, China had emerged as the third largest national art market after the United States and the United Kingdom, with 8 percent of the global share and with Hong Kong as its art capital in market terms.

- Sales of Indian art in 2007 reached 243 million euros with 70 percent of transactions concluded outside of India (in London, New York, and Dubai).

- At the edge of the economic downturn, Dubai was the leading art market in the Middle East, benefiting from the presence of the only design gallery in the region distributing a fine selection of world-class design.

- Concerning the impact of the financial crisis: while lower discretionary income will drive down luxury goods and art buying, there could be a substitution effect, with people using art as an alternative investment to stocks and bonds.

- The appearance of new markets for fine arts, in particular China, might mitigate to some extent the impact of the crisis on the global fine arts sector.

This demonstrates the market relevance of new geographies. We should however also take a purely humanist perspective. While the appearance of new art from new countries provides the hope of diversity and inclusion in larger cultural terms, the fact that all art commerce is now directed to a taste-making – if not a taste-dictating –globalized marketplace is a reason for concern. The risk is that new artists from emerging countries simply adopt or clone not just the commercial practices but also the aesthetic grammars and the philosophical visions of western art. To quite an extent, this has been the case, at least in the last 20 years. It is a complex and exciting moving picture: new art and design from new countries will have a lot to offer in terms of both aesthetics and business models.

This chapter offers five challenges to business leaders and brand strategists for them to reflect on the value of cultural diversity and the need to understand local cultures:

- To what extent do you really understand the history and culture of your local markets?

- Do you enable and facilitate the creation of new local networks of socio-cultural relevance?

- Do you proactively contribute to the creation of new cultural infrastructure to specifically help the local situation improve and progress in terms of creative excellence?

- To what extent do you play an active role in advocating cutting-edge innovation in local markets?

- Do you immerse yourself in the streets of local markets, to understand the different lifestyles that will offer you future business opportunities?

The rationale of this chapter is simple: the days of western-driven, top-down globalization are gone. We are now in the days of cultural diversity to reconcile commerce and culture. A clear understanding from a cultural viewpoint is a must for any businessperson working in a new region, be it India, China, or Japan. This chapter looks at different and diverse markets: Japan – "the mother of all Asian new markets" since the 1980s boom, China, India, and Brazil. These new geographies are reviewed not only as "markets", but also as "countries of origin" and "creative industry communities" for the new art and design of tomorrow. From whichever angle you choose to look at it, it is evident that these regions are places of unique cultural complexity. Here, there is a richness of opportunity and the diversity at its very best.

Finding 31: Reach people by truly understanding their culture

The cultural environment of fine arts and luxury markets: Japan

This "finding" looks at the cultural traits behind the taste of wealthy Japanese individuals, as these are both collectors and luxury industry customers of high relevance, and therefore, trendsetters. This is not the kind of analysis that you find in average business books: it goes further into the thinking that underpins the Japanese fine art and luxury markets. It is true now, more than ever, that in order to see the future you have to understand the past. This is why we go back in time, and get a deeper perspective on Japanese cultural history. Understanding the very roots of Asian culture is particularly important today because Asian heritage is undergoing a strong revival. What are the Japanese traits of this heritage? In order to provide an example, we will explore the notion of "beauty" in Japanese culture.

In his short treatise *In Praise of Shadows* (1935), Japanese writer Junichiro Tanizaki guides the western reader through the world of Japanese beauty as it has developed since ancient times. According to Tanizaki, the whole aesthetic sense of the Japanese is focused on the understanding and appreciation of subtle nuances and textures that are not immediate to non-Japanese eyes. Let's look at a couple of examples. Japanese taste favors ancient tools and appliances. This is considered an aristocratic sign of history, much more precious than the shining surfaces of the "new." There is more, according to Tanizaki: the intense reflex of gold in the dark is just the start to appreciating the depth of shadows, a true treasure to the Japanese eye. The shadow is more precious than the gold creating it: what a poetic notion, isn't it?

Next to the seductions of the shadows and the dark, there is the Japanese fascination for the harmonious simplicity of the everyday and its humble but spiritually superior beauty. This is not an exception in Japanese aesthetics. Writer Kazuko Okakura, in 1913, in his *Book of Tea,* invited his audience to "stop luxury, start the sublime," going beyond material possessions. In the same line of spiritual sensibility, Tanizaki writes of the "naive luxury of farmers" in describing how simple food preparation and appearance demonstrate the pervasiveness of Japanese aesthetic sophistication. The drive to perfection in such simple practices of food consumption is diffused across the entire spectrum of Japanese society, and is part of a general urge to be "impeccable" that is universally present in the Japanese way of life.

These practices originate from a tradition grounded in Oriental philosophies such as Zen and in practices such as the tea ceremony. Here the pursuit of high spiritual and moral worth in each and every move, in each and every detail is an intrinsic purpose and an objective requirement. The classic tea ceremony is today extremely popular again, among the higher strata of the Japanese population, as a way to relax and to commit free time to the aristocratic pursuit of superior beauty. This is just one indication of the more general shift towards new culturally grounded notions of luxury in the premium sector of this important trendsetting market for both lifestyle and fine arts.

These comments offer just a fragment of the rich aesthetic composition of the culture from which wealthy Japanese base of art collectors, design connoisseurs, and luxury customers come. This might suffice to justify the common view that considers the Japanese people as the world's most demanding commercial audience.

How does this historical and cultural background influence Japanese design and consumption? Years after the books by Tanizaki and Okakura, "the relentless pursuit of perfection" is a slogan that perfectly represents the mindset of the Japanese in general. Lexus used the slogan to great advantage in order to position their superior sedans against the competition. In Japanese luxury, anything less than perfection runs the risk of rejection, even if the imperfection is minute, and seemingly irrelevant to westerners. This is what we learn, for instance, in the book *Louis Vuitton Japan: The Building Of Luxury* (2004). Written by Kyojiro Hata, the president and CEO of Louis Vuitton Japan, the book describes a number of situations where products that appeared to be perfect for the Parisian consumer would be rejected by the Japanese market. As a result of the Japanese obsession with the LV brand, a simple imperfection in a stitching detail – one that is barely discernible to the naked eye – was more than enough to declare a failure to deliver the LV dream and return the product.

Such a powerful sensibility was bound not to remain isolated on the island of Japan. The next question therefore is: What is the influence of the Japanese aesthetic on global trends? We find an example in the dialog between Japanese contemporary art and world-class luxury. Here, Louis Vuitton's creative director Marc Jacobs involved Japanese artist Takashi

Murakami (see the case study in this chapter) in the design of a number of early 2000s collections of LV bags. Murakami's signature *manga*-style visuals are a household name in the art world. Once put on LV products, they generated a worldwide buzz with unprecedented impact, and created one of the biggest successes of the last decade for luxury branded artifacts. It is crucial that Murakami's art is a rigorous reference to Japan's historical tradition of painting and craftsmanship. This is where the circle is complete. The layers of sophistication which centuries of Japanese aesthetics have brought to its magnificent culture have contributed to an iconic design that has had an impact on global fashion. This has made the difference in contemporary luxury branding. Thanks to the sophisticated cultural understanding of Marc Jacobs and those at LV who pursued such an innovative match, Japanese aesthetics were "exported" from the insularity of Japan to global fashion, and from fine arts to trendsetting products, worldwide.

I hope this short cultural analysis of the roots of Japanese taste and the notion of "beauty" has demonstrated the simple need always to explore and understand our new markets. To business leaders, entrepreneurs, and the brand marketers, this translates into two key challenging questions:

- Do you have a true understanding of your local markets from a cultural viewpoint?

- What is the value that you manage to extract from the possible intercultural exchange among the different regions and countries where you operate?

The 1980s failure of major global mass marketing players to enter the Japanese market with standard marketing formats and hardly any local adaptation is testimony to the need to study in depth the history, culture, and customs of each country of operations. The study of local history and philosophies is simply homework in terms of research and strategy, because no culture is an island to be conquered and colonized any more. The next step is then to use local aesthetics and ideas to create new, surprising concept designs.

Finding 32: Reach local networks to be part of them

How daring to connect worlds apart makes the true difference: Japan and China

From Japan, our second "finding" brings us across the sea, to China. This is not easy a journey, from the viewpoint of culture, history, and trust. Wars, ideologies, and economic competition for continental leadership have played a key role across the centuries to divide and distance these two otherwise highly interdependent cultures. The commercial dialog between a Japanese art gallery and the Chinese emerging art scene that we present here should therefore be regarded as a remarkable case of relationship management and networking talent across a complex continent. In order to appreciate the magnitude of the challenge for a Japanese operator to arrive successfully in Beijing, a basic understanding of the larger Asian art market is important. The current fine arts-related relationships in this area are:

- Chinese collectors predominantly buy Chinese contemporary art, and as yet, tend not to venture into Japanese art or western art.

- European collectors voraciously buy Chinese art, and with this purchasing behavior they influence the potential future appeal for Chinese art in Japan.

- Japanese collectors, however, would not yet venture into Chinese art collecting. They stick to western art, or to traditional Japanese forms of aesthetics.

- Mainland Chinese have more fixed purchasing patterns, focusing more on Chinese art than Chinese collectors from Taiwan, Hong Kong, Singapore or the Chinese worldwide diaspora.

- Korean interactions with Japanese art started relatively later, due to the 1930s and Second World War conflicts and their postwar frictions. Korean artists however played a crucial role in the creation of important 1960s art movements in Japan, such as Mono-ha.

- In the last ten years, South Korea has become a trendsetter in Asian popular culture, with Korean television dramas and movies dictating aesthetics to important leading consumer segments across the entire region.

If we go back in time within this continental environment, Tokyo may not be the most immediately ideal location to activate pan-Asian cultural collaborations. The relative isolation of the local fine arts communities, perhaps only mitigated by exchanges with South Korean artists, was aggravated across the entire postwar period by suspicion and resentment in the Chinese and the Korean populations who were not willing to forgive the war crimes of Japanese imperialist expansion. The history of relationships among these three leading countries in the Asian continent is profound and very complex. Legacies of such a past still exist at a level of informal trust and potential prejudice. In summary, the initial handicaps for a Japanese cultural enterprise willing to enter the Chinese market are:

- the insularity of its national culture, particularly evident in the fine arts sector, with local "masters" mostly unknown to the outside world and the heavy importation of western masterpieces that started in the 1960s

- the difficult history of relationships with neighboring Asian countries, especially China, that represents an additional challenge in terms of international trust dynamics.

Given such conditions, it was remarkable what the Tokyo Gallery achieved, first by working with Korean artists a few decades ago, and second, by opening its own sister gallery in Beijing in the 1990s. This fine arts enterprise is located in a prime location, Ginza, the same district that is home to the Shiseido Gallery. The Tokyo Gallery had its first important success with the importation of western leading trendsetters in the 1960s. Such a commercial success is a major credential at historic portfolio and commercial credibility levels. This was achieved by Takashi Yamamoto, founder of the gallery. Throughout the 1960s – a decade of great economic growth in Japan – he brought to Tokyo the likes of Lucio Fontana, Yves Klein, and the masters of French informal and American abstract painting. Bringing the best of the west to Japan was a challenge in terms of customer acceptance. This challenge was met thanks to excellent relationship management, and what followed was a policy that went beyond pure commerce and artistically connected Tokyo with Korea.

Since the 1970s the Tokyo Gallery has opened the Japanese market to South Korean contemporary art – creating a true community around new Asian aesthetics – and instigated collaborations between artists from the two countries, creating new developments from shared Asian roots. This way, the gallery positioned and profiled itself not only as a very influential commercial trendsetter with great sales talent, but also as a genuine engine of cultural development with a unique Asian focus. What followed a few decades later adopted the same strategy.

Today, Yamamoto's son Hozu is a business partner in Tokyo Gallery + BTAP, and a distinguished organizer of innovative art events in collaboration with Ginza luxury retailers. He is the one who, succeeding his father, took the next challenge, by entering into the rapidly forming art market of 1990s China. Hozu's business partner and gallery co-director,

Tabata-san, operationally managed the Tokyo Gallery's interest in the Chinese art scene. Because trust between the two countries is still an issue, the first challenge was to establish solid networking assets. Also, in the mid-1990s, no real art market existed in China: it was therefore even more necessary to build strong bridges between the two Asian capitals. Tabata did so by using Chinese art expert Huang Rui (an important connection, with good professional and informal networks) as his referee and adviser. Together, they identified a suitable gallery space at low rental cost in what later became the heart of Chinese art, the Beijing 798 district. Tabata rented a loft in this Bauhaus-like industrial complex and moved on to label the new fine arts enterprise the Beijing Tokyo Art Project. This naming strategy was chosen to mitigate any perception of excessive commercialism of the Sino-Japanese gallery, because that might have created some degree of unwelcome questioning on the Chinese political and administrative side.

Once they had set up their company in the right district of Beijing, business had to be run according to the specific conditions in the field. Of course, Chinese artists and Japanese management had extremely different backgrounds. The Japanese had an in-depth sensibility based on sophisticated knowledge of western canons; the Chinese, recent children of the Cultural Revolution in China, started from a cultural void. At that time the art communities in China were populated by a kaleidoscopic generation of post-modern, insular talents, coping with a large rise in demand mostly without a solid theoretical basis. The combined operation of Tokyo Gallery + BTAP successfully linked a number of Asian artists, and greatly contributed to establish new Asian aesthetic trends in the markets of China and Japan. Even more vital, the collaborative nature of the Beijing Tokyo Art Project constitutes a primary source of value in terms of authenticity, integrity, and sense of purpose that naturally goes with commerce at its best.

Three key questions for business leaders emerge from the successful cultural connection and professional networking performed by Tokyo Gallery + BTAP:

- Is your brand an agent of positive change in an authentic debate across countries and communities?

- Is your company exploring new ways to generate value in different countries, adopting different ways of operating to fit in the local cultures?

- When it comes to China, is your operation truly integrated in the complexity of Chinese culture by means of genuine participation to communities, or is it all just about fast turnaround commerce?

Because China matters, in the next two "findings," we will focus further on China. We do so by exploring the cultural landscape of this important country in order to gain more insight into current fine arts and creative industry challenges. As a preliminary conclusion, we could say that the need to understand history and cultural mindsets (as exemplified in the case of Japanese luxury and fine arts) finds its match in the need to *be* in the field and to nurture networking relationships at a local level. I strongly believe that each local market is a different environment with different nuances that require a profound understanding, and that only the combination of deep study and an extensive presence will ensure a good chance to develop the means to bring long lasting business success.

Finding 33: Reach local markets by creating ad hoc business models

The particular environment as vital to success: China

As we all know, China has gone through one of the most dramatic spirals of growth ever seen, certainly the fastest and the most successful one ever in macroeconomic and financial terms. China has however been increasingly reminded of its heritage from the 1960s and 1970s, especially the Cultural Revolution. As discussed in our last "finding," this translated into a *tabula rasa* in matters of taste and aesthetics. In the luxury sector, it is known that some Chinese customers engage in pure status signaling; so do a number of Chinese collectors. Connections between the worlds of luxury brands and fine arts, as in the projects performed by Yang Fudong for Chanel or Ding Yi for Hermes, still represent the adaptation of a Western perspective rather than a truly Chinese approach.

A recent Salvatore Ferragamo exhibition at the Museum of Contemporary Art in Shanghai showed the limits of both the curatorial team and the maintenance staff of the museum: they could not effectively display and efficiently maintain the sophisticated applied arts items on show. There is a historical great past of Chinese splendor in fine arts and luxury, and there is no doubt that China will take on a leading role in international culture and the arts in the future, on a regional basis to start with, then on a larger scale. The question in this "finding" is however: How will this happen? This "finding" (and the next one) reviews the way

China has evolved in the last decade or so, with particular focus on the most cutting-edge, experimental sectors of contemporary art and applied arts. The biggest question for business leaders and entrepreneurs is challenging as usual: Do you generate innovative business models with a longer-term perspective, for the best benefit of the region where you operate?

In China more than anywhere else, it is local networking that makes the difference. Every businessperson who has the opportunity to become familiar with the Chinese way of doing business knows the concept of *Guanxi*. This is a word that defines the networks of relationships within a social circle, where reciprocity and solidarity are regulated by a compensatory system of mutual support. Thanks to *Guanxi*, no man or woman is an island, in China – provided they use networks and access to return favors when required. *Guanxi* in China is the social capital that Tokyo Gallery so needed, and built up so well, to ensure the success of its Beijing operations. The concept of *Guanxi* exemplifies how China is not a country where business practices at an informal level can be performed according to pure contractual and formal agreements. Understanding the deeper layer of social and cultural interactions that regulate the practical side of Chinese communal life is a necessity. If you are not Chinese, becoming part of this complex system of human interdependencies is a rather unique privilege.

Achieving unique levels of *Guanxi* – and from there creating new ways of doing business – is what Italian cultural sector and creative industry leader, Davide Quadrio, accomplished in nearly 20 years of residence in Shanghai. Quadrio arrived in Shanghai in the 1990s, when there was no contemporary art scene in China. He was fluent in the Mandarin language. He developed a street-smart practical knowledge in terms of understanding the complexity of public sector affairs in the city. On this basis, Quadrio created his unique venture, BizArt. The word "BizArt" describes the connection between business and art. This is the DNA of Quadrio's business model: revenues in the applied arts sector (design, advertising, brand consulting services) were reinvested into the creation, production, and management of up to 50 fine arts projects a year. Through the years, BizArt has been one of the key drivers in the creation of the Shanghai contemporary art scene. In parallel, BizArt branched out into consulting services, with Quadrio taking up projects with Droog Design in the strategic corporate branding domain, and playing a vital role in the Shanghai luxury sector, with his creative direction in the world-class Bund 18 creative center. In time, BizArt became an institution increasingly specializing in the realm of public art, with a strong research component and international ties at the highest level, from the prestigious Prins Claus Foundation in the Netherlands to the Van Abbemuseum.

On one hand, we could conclude that BizArt developed an original business model while creating to some extent a new art scene, connecting it to the new-to-China domains of marketing, design, and urban design. On the other hand, the local dimension of BizArt magnified its structural weaknesses, such as the necessity to stabilize the business model in the face of complexities. Of course, this was not an easy task, especially given a number of structural shortcomings within the Chinese system. For example there is:

- a legal system not geared to optimal western standard levels of efficiency and transparency

- a predominance of financial control exercised by powerful local lobbyists

- a worrying drift in terms of ethics, as started and stimulated by the economic boom in the last 20 years.

Perhaps the most important advice that Quadrio's BizArt experience could offer to corporate leaders and business entrepreneurs aiming at success in China is to take things slowly. A similar invitation to take time and do things at the correct speed when approaching China as a new market was also one of the key recommendations by Chinese tycoon and luxury market expert, Dr Richard Lee, in a recent keynote stage interview at the ESOMAR APAC conference in Beijing. Founder of the China Lifestyle Premium Enterprise, Dr Lee imported the first Ferrari ever to contemporary China and is still today the official importer of Ferrari Maserati into China's major cities. What conclusion can we derive for business readers? If you want to do serious business in China, then be serious about China. Go to China with a willingness to truly understand the field. Think in the long term and create solid relationships over time. Invest in the local environment to truly make a difference. And, of course, develop as much *Guanxi* as you can, all along the way.

Finding 34: Reach and nurture local creativity circles

Move beyond plain profit making, to generate long-term value in the local environment

Creativity is one of the vital assets to be planted, nurtured, and harvested to find solutions to the challenges we face today – challenges we have never experienced before in our lives. In Chapter 3, we discussed creativity as a faculty of individual humans, either in isolation or in social interaction settings. In this "finding," we explore the impact of the larger environment of Chinese recent national history and culture on creativity output. The underlying message for business leaders and entrepreneurs is very simple: move on from the limited vision of countries like China as an outsourcing production facility, and engage in generating true value for local individuals and communities in the longer term, because it is such long-lasting value creation that will ultimately entrench and embed an enterprise or a brand in any market. To do so, a profound understanding of Chinese creative industry circles is a necessity, not just an option. The key question to challenge business leaders and corporate management is: Does your enterprise advocate a culture of creativity in the local environment where it operates?

Of course, this question applies and relates, above all, to the work of (Chinese) national and local policy makers and public officers. These are the professionals who are formally tasked with the definition of next steps. The environment in China is such that creativity is greatly needed for progress: this is where fine arts and design rule by default, and this is why fine arts and design are expected to help greatly in this challenge. The approach taken by the Chinese public leaders was to endorse the erection of creative clusters in urban areas of Chinese leading cities, with sector 798 in Beijing and its equivalent in Shanghai, M50, as the vital focuses. The ambition was to see creative quality emerge from fine arts and spread to industry and society with dynamism. Likewise, the investments in applied arts over the last decade were no less impressive, with some 400 design academies being created in recent years. Nevertheless, creativity is not an easy currency to find in China, and high-quality creative output is even rarer. Why? As we discussed in Chapter 3 for advertising and design in general, training is vital in this environment, and academic education even more so. Chinese didactic systems and approaches however present a number of handicaps for the growth of a true creative edge:

- The education system is geared in a circular mode, reinforcing and reinstating the traditional structures, without genuinely starting and stimulating individual creative development.

- The expectations of political leaders are that whatever manifestation of applied or fine arts will be generated in the cultural system, they will be harmonious and aligned with the general purposes and goals of Chinese society towards economic growth and prosperity.

- Within Chinese social systems, the level of tolerance towards deviants and contrarians is remarkably low: the main focus is on integration and productivity, with a great degree of uniformity. The result is a lack of the critical thinking that helps the growth of lateral thinking and creative excellence.

As an example of the stalling within Chinese creative industry, we can report the disappointing evolution of the Chinese cartoon industry. This seemed an ideal place to plan the next successful steps of Chinese cultural production: relatively cheap with respect to blockbuster features, with an established creative process defined by a few decades of existence, and tentatively delivering the kind of Disney storytelling that a society constantly aiming for inner harmony might naturally benefit from. For these reasons, major investments were granted with the ambition of producing a next generation of world-class *manga* cartoons, made in China and intended for worldwide distribution, with great hopes of economic success. However, no major production resulted from this public-sector-funded strategy. The reason could lie in the drive we mentioned towards social harmony, undermining the sharp alternative viewpoints that are required to truly make a difference in the creative process. This holds true even when the creative process is all about inventing the alternative worlds of successful family-appeal cartoons for the silver screen.

If the challenges are clear, what are the solutions? The next step towards China emerging as new world creativity leader could be a change in terms of performance indicators and actual measurement of success: from sheer quantity of stuff produced to quality of experiences and of products, from the quick comfort of profits to a new excellence from the point of view of aesthetics and semantics. Ultimately, a deep change in the educational system will be needed, reflecting a different role of mentorship and advocacy by professors towards students. This would intervene at the center of the circular system. The purpose should be to break the loop, introducing a whole new mindset to advocate creative leadership.

It might look as if the challenges described so far belong to the sphere and scale of public offices and political leaders. This is not the case, and the demonstration of this is offered by Quadrio's work. His recently launched Bangkok-based enterprise, ArtHub, has the specific mission of connecting the members of the Asian creative community and bringing Pan-Asian fine arts to a whole new level of quality. This is once again an important asset in terms of *Guanxi* and networking across the continent. Offering unique bases for critical thinking and advanced curatorial support, both BizArt and ArtHub demonstrate how individual enterprise can still make the difference in terms of innovating and pushing the boundaries, by enabling and facilitating, even in the challenge of pursuing creative excellence in China. Quadrio has proved that one person can make a difference, as long as the vision is clear and the direction is coherent over time.

The natural question that arises as a conclusion to this "finding" is: What could business leaders of large corporations do to contribute to the rebirth of "creative China"? If just one talented foreign maverick like Quadrio achieved so much by starting and staying independent, where were the *Fortune 500* brands in this process of Chinese creative renaissance? Here, the message to corporate managers and brand marketers is a simple call for action: just like Quadrio did, get your hands in the dirt. Instead of sitting in your corporate ivory towers, go in the streets and make things happen from the bottom up. The time of pure deployment of Anglo-Saxon models in business and branding is coming to its end: it is time to rediscover the world from the point of view of local cultures. The space to make things happen is phenomenal. And this is true not only for China, India, and Brazil: any new, emerging market, will offer huge opportunities to those who are there to listen, learn, and contribute true value over time.

Finding 35: Forget high and low: reach culture at all levels

From the elite to the streets, from India to Brazil

In this chapter, we have highlighted evolution in Japanese and Chinese fine arts. We mostly focused our attention on the high-art and luxury product categories. In this "finding," we move our attention to two emerging economies of great relevance, India and Brazil. We explore local dynamics in the distinctive fields of fine arts and of street culture. Two cases form the backbone to this chapter, offering an overview of two opposite experiences:

- The scouting and nurturing of high art: How an important leading gallery achieved, over the years, the profiling and exposure at world-class international level of contemporary Indian artists.

- The connecting of popular culture to market research: How a grassroots foundation used creativity to produce social innovation in the under-privileged streets of Rio de Janeiro, while creating an innovative market research offer of ethnographic studies leading to potential new business opportunities.

India is one of the darlings of the contemporary art market. Indian art is one of the aesthetic vernaculars supported by Charles Saatchi, and so a relative explosion of interest in the global collecting base was inevitable. The Indian art scene is currently in a state of profound transformation, as is Indian society. The elevation of Indian artists to celebrity status is a relatively new phenomenon here, but a rapidly spreading one. In parallel with such dynamics of success, the emergence of solid design schools and the diffusion of new creative techniques such as computer graphics, photography, tattoo, and textile design, are promising fruitful multidisciplinary developments. Within this environment, Gallery Chemould, recently renamed Chemould Prescott Road, is a major player at many levels: commercial, curatorial, and cultural, among others.

Founded in 1963 by Kekoo and Khorshed Gandhy, it rapidly began to represent some of the best Indian artists at international level. The gallery is a regular exhibitor at important regional and national fairs, from FIAC of Paris to ShContemporary in Shanghai. This enables visibility among international collectors, with the added benefit of positively presenting Indian art in general. The latter has always been an explicit goal of the gallery. It has been pursued through collaborations with important museums worldwide, starting from the 1960s group show "Art Now in India" which traveled in Britain, Switzerland, and Germany. Such an active role at the highest level of cultural prestige also includes collaborations with the likes of Documenta, the important exhibition event that occurs every four years in Kassel in Germany. The organization of these shows was a strategy for an art commercial enterprise operating in an environment where no structural investments are made by the public bodies. Additionally, the creation of its own publishing house followed, to pursue the diffusion of catalogues and books on Indian art and artists.

Chemould Prescott Road shows an ability to take the lead and engage in a promotional strategy geared to the higher streams of cultural presence. The representation policy (including artists from all generations) and the curatorial policy (offering the opportunity to experiment with new media, new techniques, and new formats) enable this gallery to position itself as an active player in the future development of Indian art.

Two questions emerge from the inspiring experience by a leading gallery of India in the making of Indian fine arts. First, Does your enterprise act as a creator, facilitator, and stimulator of excellence within its industry and category? This is the approach chosen by Chemould Prescott Road. The gallery emerged by enabling the success of Indian art at the level of high-art world-class circuits and circles. The conclusion for corporate managers and business leaders is rather obvious: participate, participate, participate. Nothing is more valuable in terms of longer-term relationships than a continued presence in the field, advocating culture and stimulating quality to emerge generally within local societies.

The second question that emerges is: Does your enterprise perform in-depth (market) research into the "real world" of developing trends, where younger generations are creating the future of the local culture? Of course, we could correctly state that Chemould Prescott Road does so by investing in the art of new generations and by maintaining a primary research profile in spite of its international success. To truly address this second question, however, we move to Brazil, where we follow the strategies adopted by Stichting Caramundo to enable viable business modeling and true emancipation for street art talents and underprivileged youth.

We first met Caramundo in Chapter 4. Caramundo collaborates with institutions such as the Nederlands Fotomuseum and the City of Rotterdam for important cultural programs, and focuses on popular culture in metropolitan Latin America. Here, art expressions do not come according to a canon, a genre, or a technique. The scope of Caramundo's activities covers fine arts, independent cinema, music, and culture, as well as the sociological dynamics of street life in the *favelas* and peripheries of cities like Sao Paolo and Rio de Janeiro.

Parallel with its main business process, Caramundo offers important opportunities to corporations in terms of research projects. Deeply entrenched and optimally integrated in the world-famed *favelas*, areas of great danger to outsiders, Caramundo is led by an academically sound anthropologist who has the reach and the capability to set up ethnographic research programs. These include direct onsite visits for marketers, designers, and brand managers. This way, Western executives have the privilege of experiencing at first hand the life of underprivileged areas. Here, they can meet face-to-face with prospective customers for tomorrow's new propositions, and truly understand their needs and their local environment first hand. This is a precious chance to understand lifestyles an ocean away from European corporate ivory towers. This opportunity to be there personally, in the very place where street culture happens, is exceptional in its experiential value. This is not a theoretical market seen from the sterilized lenses of statistic sampling: this is real life, presented at the crossroads where new aesthetic trends are being nurtured by their creators.

We can conclude that Caramundo and BizArt cross-reference each other from at least three specific points of view:

- acting as insiders in new worlds in the making, with the participation of the involved activist driven by a precise vision

- connecting with difficult realities, earning *Guanxi* (or the equivalent in Brazilian *favelas* in social capital terms) and promoting the next wave of local talents by sharing, enabling, and advocating

- offering the outside world the opportunity to access the local environment in an unique way: not as scientists studying a subject but fully integrated ambassadors, connecting research commissioners and local people.

It is important to reiterate that Caramundo does not do charity: Caramundo does business in terms of emancipating the underprivileged from poverty through opportunities, and plugging corporate research straight into the streets of Brazil. The business model embodied (with vaguely colonialist paternalism) by the standard programs of corporate philanthropy addresses local communities as passive beneficiaries. As we mentioned in Chapter 4, Caramundo aims instead at creating a deeper connection through a real presence and true compassion, which result in far more benefits in the longer term for all stakeholders involved.

What ultimately is at stake here is the actual role of global enterprises in local markets. In concluding this, we complete this chapter with the fundamental questions that result from the different experiences of Chemould Prescott Road in the high art and Caramundo in the popular culture of their countries. These are challenges to be considered by business leaders and corporate managers:

- Does your enterprise undertake research to understand the local environment across different levels, from the cultural elite to the underprivileged masses?

- Is your company in China or in Brazil to just study, design, and sell stuff, or is it genuinely willing to participate in local culture, and its making, with an active and proactive role?

The main conclusion of this "finding," is that the world is not a colony of the west any more: it is time that advanced economic enterprises learned to look at the world in a different way, with the genuine aim of initiating a true debate on an equal basis. Never before have advanced economies and emerging markets been so much in need of each other. The last and perhaps most important question is: Who will take the lead in changing the rules of the globalization game as we know them so far?

Wrapping up: from new geographies to new ways of doing business

To round off this chapter, we refer to an art show where distant cultures ideally met in a spiritual fashion. In 2008, an Aboriginal art master, Emily Kame Kngwarreye, had a major retrospective at the National Art Center of Tokyo, entitled "Utopia." This happened in the same city where the highest concentration of luxury brands is, at the same time when Chanel was operating its Mobile Art Project (short-circuiting art with consumption). I saw myself how this show enchanted Japanese audiences. It offered the opportunity to ultra-sophisticated Tokyo museumgoers to reconnect to the core anthropological representation of a true tribal artist.

A true amateur from a western viewpoint, Kame Kngwarreye never attended an art school or had a formal education. She started to paint only in later life, after turning 70 years old. In pursuing such a late step, she was supported by a social program of her regional government in Australia, but this was limited to technical advice. Other than this basic support, her aesthetics represent purely and entirely her own shamanic connection between her earth and her people. It must be specified that, beyond aesthetics, through her art and within her tribe, Emily played a specific role of formal authority in an ancestral cultural structure.

Divided into five different styles with aesthetic cohesion and consistency, Emily's works are not those of an amateur to the eyes of curators and art professionals. "Utopia" traveled next to the National Museum of Australia in Canberra. The highest price reached for Emily's work was US$1,000,000 at New York auctions in 2007. Nevertheless, it is not just her recent market status that truly matters here; what matters is the opinion of those in the know in the fine arts system about Emily's acknowledged value, and it is important. It should be stressed that she belonged to the world of amateur artists almost until her death: just like one of Caramundo's street artists, she was painting to express what she believed needed to be said within the scope of her tribal role. From this fundamental low-point of artistic amateur production, Emily's oeuvre has been elevated to the highest degree of high-culture value for first world museum-goers. The ecstatic reaction of Japanese audiences when exposed to the large canvasses of powerful color combinations was the best nonverbal statement of the value of Emily's work and of true art in general terms. We regard this as an affirmation of the universal power of fine arts. Emily's work is also the most direct proof that religious practices are still anthropologically connected to the ultimate meaning of art, and vice versa.

As I said in earlier chapters about museums, there is a remote link in western culture between fine arts and the spiritual elevation of religion through beauty: although it might remain only in the remote collective memory of each of us westerners, this is an undeniable trait. Of course, analogies and metaphors connecting fine arts to specific realms of spirituality are always culturally specific, and therefore differ from region to region.

Stressing the general relevance of such cultural understanding was the purpose of this chapter. Here, we discussed the regional texture of differences that lie below the surface of global markets of arts and culture. Our aim was to connect this texture with business-related challenges and questions of universal validity. If these challenges are local, the ultimate scope of this chapter is global because we highlighted new insights and ideas that will be necessary to grow and prosper in tomorrow's changing world.

Superflat: Takashi Murakami and Japanese aesthetics

Few other contemporary artists have acquired the celebrity status that Takashi Murakami had in the 1990s and 2000s. Having developed a signature style and leading an enterprise, the Hiropon Factory, which has some similarity with Damien Hirst's production capability and Andy Warhol's Factory, Murakami became even better known because of his connection with Louis Vuitton. As presented in our first "finding" in this chapter, Murakami's signature-designed luxury products could appear to be frivolous to art lovers with a taste for philosophical art, and ephemerally contemporary for those who prefer art to reconnect with tradition. Murakami's work is however neither shallow nor lacking historical perspective or context. As captured in his 2000/01 book, *Superflat*, the artist has a clear vision at the synchronic level for our everyday and at the diachronic level for history.

In terms of the present everyday, Murakami is as a super-post-modern artist, working in an apparent vacuum of ideology that connects the storytelling lines of *manga* comics with telephone sex services. In reality, as much as Murakami's vision appears nonpolitical, his vision of superflat art – his own art – is deeply historical and profoundly Japanese. Murakami starts his ideological platform within the "Superflat Manifesto" with a philological clarification: while the words "entertainment" and "profession" match clear concepts in the Japanese language, the world "art" is a foreign import from the west, and its actual mental image lacks any focus or precision in the mind of Japanese artists. From this starting point, Murakami reconnects and almost rewrites the lineage of Japanese aesthetics. He does so in a highly impacting visual sequence that connects 1970s *manga* and 1990s MTV cartoons with classic paintings established in the national tradition for centuries. Such a visual stream connects popular culture icons such as 1999 groovisions and photographer Hiromix with contemporary emerging artists such as Yoshitomo Nara. This is then mixed with ancient masters of tradition such as seventeenth-century Kano Sansetsu and eighteenth-century Ito Jakuchu. Murakami's "superflat" provides a different vision of Japan, one that perhaps is not philologically correct, but is surely stimulating in terms of connecting the banality of the everyday with the sophisticated traditions of the past.

What Murakami achieved on a "vertical" level, by reconnecting Japanese fine arts history to its roots and to what he considers its future, is mirrored by his marriage of luxury fashion with his fine arts signature icons. The circle is complete: it is business-wise one of the top successes of the last decade; it is aesthetically central to global art; it is sociologically an acute insight and a perfect mirror of contemporary Japan and perhaps, to some limited extent, of a more general Asian environment.

CHAPTER 9

The future of fine arts – according to public opinion

So far, we have followed a path of 35 "findings" as derived from qualitative research, structured interviews with thought leaders, and movers and shakers in the world of fine arts. The purpose of this has been to gain insight and inspiration from the creative and cultural sectors for the benefit of business leaders and brand marketers, including market research professionals.

In this chapter we change perspective, using a sophisticated quantitative market research methodology to identify future patterns of potential evolution in the world of fine arts. To complete this change of perspective, our respondents change from opinion leaders and professional trendsetters in the world of fine arts to ordinary people. This chapter is designed to provide a different viewpoint for the main benefit of museum managers and fine arts business leaders. After providing learning from the cultural sector, our market research now enables them to reflect on their own future. This chapter was created in collaboration with Peanut Labs, the developers of the Sample 3.0 methodology, and with special thanks to the generous contribution of Moskowitz Jacobs, Inc of White Plains, New York, in particular Alex Gofman, CTO and VP, who conducted the primary analysis of the data.

The business question behind this chapter

What is the future of fine arts and "design for collectors" according to a statistically chosen sample of ordinary people?

An introduction to the findings in this chapter

Our journey across *The Golden Crossroads* of fine arts and design has brought us from the depth of psychoanalysis to the new emerging regions of future growth. We followed the insights of thought leaders in the fine arts and design, and we benefited from their vision, ideas, and intuition to gather "findings" in order to inspire business leaders and brand marketers. It is now time to engage with one of the most advanced quantitative market research techniques. We will do this in order to offer a glimpse of the future of fine arts and design to those who will make it happen: artists, collectors, museum managers, and art dealers. Creative and cultural industries professionals now have their own opportunity to gain new insights into their own business from *The Golden Crossroads*'s multidisciplinary approach.

The research methodology in this last chapter is Rule Development Experimentation (RDE), and its world-class background is explained in the case study. In collaboration with Peanut Labs, the sample provider, Alex Gofman applied RDE principles to the world of fine arts, and the analysis of its future. We decided to measure the opinions of the British and Americans as these are still the two countries that drive markets in terms of fairs, auctions, galleries, media, and magazines. If China or India could be the places of tomorrow's fine arts, today's predominant opinions are still largely formed in New York, in London, and in other cities in these two nations. The experimental research was conducted in spring 2009, with an outlook over the next five years. The sample was structured as follows:

Total: 418

Male: 206

Female: 212

United Kingdom: 201

United States: 217

The sample was not selected from gallery goers: it was made up of a larger base from the populations of the United States and United Kingdom, composed of people who might go to museums or even perhaps buy art, if properly approached by curators and galleries. The sample comprised a complete spectrum of occupations: full-time students, unemployed people, part-time and full-time workers, and retired people, with space for other options to be separately accounted for. The age groups were:

18–29 years

30–39 years

40–49 years

50+ years

The quantitative measurement was first aimed at identifying the expected dynamism in the future developments of fine arts and design for collectors. This is statistically done by measuring the baseline, or the percentage of respondents who perceive that the field of analysis is in a state of dynamic change. As Gofman explained, a high baseline means movement in the analyzed area. This first data gave a very clear feedback: nearly one quarter of the entire sample thought that the art world is going through change. The overall baseline was 24 (meaning 24 percent of respondents), but this varied considerably between components:

Male respondents: 28

Female respondents: 19

UK respondents: 14

US respondents: 33

Clearly male respondents see the world of fine arts as much more apt to change in the near future than women do, and US residents think so much more than UK respondents. The crucial question is: Where is the world of fine arts going, exactly?

Starting from this framework of RDE data, the rest of this chapter tries to give a feeling for the different directions of change in the next five years. This chapter offers three "findings:"

- An overview of the general direction extracted from the qualitative expert interviews and fed as primary input into the RDE testing process.

- The schematic description of two market research segments defined from the data distilled by RDE. These two segments explore two different mindsets that aggregate US and UK public opinion on the topic.

- Additional detail and comments on the different information that the RDE data provided at the level of analysis (by gender, by countries, and by specific age group), to give more depth.

By the end of this chapter, we will have added to our earlier considerations and explorations the validation that only the science of statistics can offer, when used in the right way. We will not dwell on the numbers, as this is not our purpose: we will instead try to derive inspiration and insights from the numbers, to build a different viewpoint on our mosaic about fine arts and design.

Finding 36: General directions for the future of fine arts and "design for collectors"

Extracting the essence from the qualitative wisdom of experts

RDE organizes the input to sample respondents in a number of conceptual "silos:" this is where individual lines of text – statements (or "items") – are extracted and assembled in onscreen vignettes for the interviewees to rate. For this study the six silos were defined on the basis of the expert opinions aggregated from the qualitative interviews (discussed in the remainder of this book). These clusters offer some initial conclusions about the future of the fine arts systems as seen by the contributors. The organizing principle of the RDE silos not surprisingly respected the flow of our journey so far, using a sharply focused composition of our "findings" into the hypotheses to be tested through quantitative interviews.

The six conceptual clusters were organized in six lines each, selecting key themes and translating them into simple, one-sided, practical statements for the sample panel to give their opinion on. For example, we did not ask a question about DesignArt, as this is a technical definition for art specialists, but used the term "design for collectors" so the point would be readily understood. (Chapter 1 outlines the step-by-step process of RDE.) What was the overall picture offered by the six conceptual silos?

1. Fine arts and the artist

In the first cluster, we tested six statements related to the nature of the connection between fine arts and the actual protagonists – the artists. As one of our qualitative research contributors put it, the future of fine arts lies with artists because they and only they are the ultimate custodians of what will be produced. Did the public think so too? These were the RDE statements we used to investigate this:

1.1 Fine arts are the production of an artist as acknowledged by the mass media.

1.2 Fine arts are what trained academy graduates do with paintings, sculpture, and media.

1.3 Fine arts are anything produced by any self-proclaimed artist.

1.4 Fine arts are always connected with the social success of the artist.

1.5 Fine arts are always the creation of an outcast or a dropout.

1.6 Fine arts are the manifestation of deep personal processes by an artist.

2. Fine arts and art movements

The second cluster followed the book chapter sequence, from artist to art movement. What is an art movement and what do people see as its ultimate purpose? We asked our respondents to rate these statements:

2.1 Art movements define themselves by means of their manifestos.

2.2 Art movements are the outcome of a loose group of artists.

2.3 Art movements are more relevant than individuals.

2.4 Art movements are necessary because the art world operates according to the principles of branding.

2.5 Art movements discipline and streamline the power of creativity.

2.6 Art movements are just an accidental clustering of individuals, done by critics.

3. Fine arts and art communities

Next was the theme of communities and collectors (the subject of Chapter 5). For the purpose of this statistical exercise, we tested and measured these issues through two clusters of statements, one dedicated to communities and one to collectors. The community-oriented statements were:

3.1 Art communities are crucial to the quality of city life.

3.2 Art communities should be supported by means of public funding.

3.3 Art communities are a superfluous luxury in times of crisis.

3.4 Art communities are crucial for a city to enjoy business success.

3.5 Art communities are the cultural research labs of advanced economies.

3.6 Art communities connect fine arts with design, advertising, and other creative industry business.

4. Fine arts and collectors

This complementary silo looked at social and business aspects of collection. It included a first "checkpoint" about the short-term sustainability of the art market. The statements were:

4.1 Collectors are exceptional individuals who support creativity and self-expression.

4.2 Collectors are yet another example of personal greed over social interest.

4.3 Collectors are masters of business with a sense for art – they can teach us a lot.

4.4 Collectors should advise business enterprises on the basis of their experience in fine arts.

4.5 Collectors are not relevant to determine the future of fine arts as the public sector will determine what is next.

4.6 Collectors are part of a fine arts market bubble: this market bubble is set to burst.

5. Fine arts and museums

The theme of museums is of course central to an analysis of the future of fine arts. The statements in this silo had two broad purposes. The first was to ascertain the perceived value and role of museums. Second, in line with Chapter 6, we moved on to explore the potential connections between museums and business. The statements were:

5.1 Museums are an essential backbone of our culture.

5.2 Museums are a luxury we cannot afford to pay for in a time of crisis.

5.3 Public museums are science and knowledge; private museums are just a hobby.

5.4 Museums are a vital source of research for business enterprises.

5.5 Museums are a source for designers and marketers to rethink their practices.

5.6 Museums should be private to benefit from the best business management talent and principles.

6. Fine arts and design beyond today

For the sixth and last conceptual silo, we selected a few options from Chapters 7 and 8 on design and new regions. This cluster offered us the opportunity to test a selection of statements that could help professionals in the fine arts and design systems get a sense of where people believe the system is going. The statements were:

6.1 The future of fine arts lies with urban design and city beautification.

6.2 In five years, "design for collectors" will be exhibited and traded as fine arts.

6.3 The future of fine arts lies with Asian artists and curators.

6.4 The future of fine arts lies with South American artists and curators.

6.5 The future of fine arts lies with Indian artists and curators.

6.6 In a time of crisis, the fine arts system (as we know it) will simply dissolve.

While these conceptual silos show how we structured the research, that is not how the respondents were exposed to it. They were asked one core question, repeatedly: "How well does this 'concept' describe the direction in which you feel that the world of 'fine arts' and 'design for collectors' is going?" Each time this was asked, they were shown a different combination of four of the 36 statements. A feature of this methodology is the mixing of different lines from different clusters into one vignette. For example, a respondent might have been given these statements together to appraise:

1. Art communities should be supported by means of public funding.

2. Collectors are exceptional individuals who support creativity and expression.

3. Museums are an essential backbone to our culture.

4. In five years, "design for collectors" will be exhibited and traded as fine arts.

Respondents rated each statement on a scale from 1 (equivalent to: this combination answers the research question very poorly) to 9 (equivalent to: this combination describes the research field perfectly). Moskowitz Jacobs, Inc.'s proprietary algorithms then managed the collective response and processed it into coherent and relevant information, to be interpreted, analyzed, and clustered into segments, as presented in our next finding. These segments give us the best synthesis of where public opinion is that the cultural and applied arts sectors are actually going.

Finding 37: People's statistical preferences, explained

US and UK public opinion on the future of fine arts organized into coherent segments

This "finding" presents a statistically sound segmentation of the opinion of our US and UK respondents. We identified one main dividing line of opinion about the future of fine arts and design. The segments were clustered and named to represent their key focus: "Future art as autonomous artifacts" and "Future art as social process." Let's look at them in more detail.

Segment 1: Future art as autonomous artifacts

Art "objects" have often been described as "autonomous" because their purpose is not functional. Such artifacts exist for reasons of aesthetics or pure beauty, not to perform a practical task. Design is different: a chair is primarily for sitting in. Nevertheless, DesignArt brought design closer to the notion of style, to the point where some have labeled its products as "autonomous design," in order to underline their close affinity with fine arts. This segment subscribed to this idea as the future direction of development for the field of fine arts. The notion of the artist as *the* representative of the fine arts is secondary here: what really matters is just art in its highest form. This holds true to the extent that the social success of artists was, perhaps surprisingly, rated as irrelevant to the future of fine arts. At the same time, this segment perceived larger art communities as extremely important – even to the extent of being supported by public funding, although not for the direct purpose of increasing the quality of city life. This is a vital segment in terms of the richness and the complexity of its feedback to the survey.

When it comes to the connections between fine arts and business, the position of these respondents is quite thought-provoking. They saw both collectors and museums as extremely important in terms of potential learning opportunities for enterprise leaders and brand marketers, to the extent that they will be the sources for professionals to rethink and retrain in their ways of doing business. Here, the art market is definitely seen as a positive place of inspiration. In particular, the future development of marketing and design talent will be positively pursued thanks to fine arts sources. Perhaps the fact that fine arts are autonomous qualifies them to be an alternative space to reformulate business visions and to review business frameworks. One detail: this segment also gave a very specific negative response to the hypothesis that South America might be the hothouse for future trends in fine arts. Given the high levels of social concern in that region, could this be confirmation that the social dimension of future fine arts is not vital to these respondents?

The analysis located 225 respondents (that is, a small majority of all respondents) in this segment. However, these were not the respondents who gave the strongest responses in terms of baseline value, particularly in comparison with Segment 2, and they did not react very strongly to the vignettes that we had set up (that is, they rarely opted for the extreme 1 or 9 values). The analysis identified their responses to only six statements out of the total of 36 as relevant to constructing an aggregated opinion.

Segment 2: Future art as social process

This second segment included 193 respondents out of the total sample of 418. Their levels of opinion and participation were much higher than Segment 1's. For example, their highest rated statement was statistically weighted at more than twice the highest rated statement within Segment 1. Their response suggested that the future of fine arts and design will be definitely more socially attuned. This reflects the general opinion of the experts and thought leaders interviewed for our qualitative research study – with subtle shades of difference, though.

These respondents reckoned that artists as individuals will continue to be – or regain their role as – the driver of fine arts. Unlike Segment 1 respondents, they positively appraised the social dimension of success for artists. However, that does not mean they saw the future of fine arts as focused on a separate class of "geniuses." Although they saw future fine arts as strongly associated with the personal creative processes of artists, this did not seem to mean artists operating in solitude, according to their own devices, or without any proper training. The validating role of art academies in training artists was a strong part of the picture, and there was a clear focus on art movements as the most relevant vectors – well above individual contributors. Additionally, they saw the role of art critics and of manifestos in the creation of art movements as critical. This added up to a strong picture of a socializing and socialized future for fine arts, with its focus on the appraisal of the crucial role of tomorrow's art communities.

The real focus of this segment is on the very backbone of fine arts, in the past as much as (seemingly) in the future – communities. This segment reacted positively to most statements related to art communities. They saw art communities as crucial to the future of cities in terms of quality of life and business opportunities for growth. They were therefore extremely concerned that the existence of art communities might be challenged as a superfluous luxury in the current economic crisis, and advocated public funding to protect and nurture them, just as Segment 1 respondents did. To a lesser extent, but still with comparable levels of interest to Segment 1, these respondents saw art communities as the connecting factor between fine arts and applied art: that is, design, advertising, or any other creative industry business. In this respect the role of art communities as the cultural R&D labs of advanced economies was fully acknowledged and supported. Museums are no exception to this overall positive perception of future fine arts as social assets. They were likewise seen as important backbones not just of our business wisdom but of our entire culture. Collectors however were not appreciated as future patrons of the arts but were instead seen as individualists not acting in accordance with society's interest. Could the most recent associations of high-end fine arts and design with the economic situation have played a role here?

The rich texture of this segment is confirmed by yet another feature – their view of design. To these people, design in the future will continue to be traded as fine arts: for them, the transition of design from applied art to fine art will inevitably be complete by 2014. To this segment, there is no contradiction between a social vision of fine arts and an artistic vision of future design. This leaves a great deal of hope that – beyond the current crisis – a next positive phase of "engaged design" will continue to be appreciated and collected. It would not be the first time, from the Old Masters of painting to conceptual art or Arte Povera, that the identities of fine arts and design are being redefined. This time it will certainly be for the better.

Preliminary conclusions on fine arts futures segmentation

In conclusion, it is important to reiterate that these two segments capture the essence of the RDE statistical analysis. It was clear from the research that there is a divide between two different positions in US and UK public opinion. The segmentation offered some clear ideas about where the world of fine arts and design is heading, and I feel this part of the research was very much complementary to the opinions derived from the one-to-one interviews with thought leaders worldwide.

However, no segmentation can retain all the richness and granularity of individual responses. Let us move to look at some other information derived from the full set of data, which will help to enrich our understanding further.

Finding 38: A detail is worth a thousand segments

Digging into a statistical goldmine to find out more about future fine arts

This third and final "finding" offers fine arts and design professionals different insights and nuances extracted from the response by our sample audience to the statements put to them. We explore different insights as defined and clustered across three key dimensions:

- country: United Kingdom versus United States

- gender: male versus female

- age bands.

Country analysis

Both national groups of respondents agreed on one specific proposition: art communities should be supported by public funding, with US respondents displaying a slightly more positive attitude than UK ones. US respondents indicated that art movements will be vital in helping creative talent. UK respondents endorsed museums as vital, both as cultural institutions and as a source of inspiration and change for designers and other creative industry players. We should bear in mind here the long tradition of British museums performing an educational function.

In more detail, North Americans indicated that they do not expect South America to take the lead as *the* next hub of fine arts. This directly goes against one of our contributing thought leaders, and might seem surprising, given the growth of Hispanic audiences in the States, and the fact that North America is an obvious key market for South American artists. It will be interesting to find out over time which view was right.

Gender analysis

The 206 males and 212 females were spread across the spectrums of age, occupation and education. Alex Gofman's preliminary analysis showed that on average, male respondents tended to rate the statements provided about 10 percent more positively than female respondents did. There were a number of individual gender differences in response to the statements, too:

- Male respondents saw fine art movements as necessary to operate the art sector on the basis of working branding principles, while female respondents saw movements as improving the efficiency of frameworks for creativity which are self-organized thanks to manifestos.

- Female respondents typically saw art communities as connecting the various branches of the creative industry, and therefore strongly supported their public funding. The impression that art communities could become an unaffordable luxury was stronger among female respondents, by about 6 percent.

- Male respondents saw future art collectors as purely driven by self-interest, while the female respondents were neutral about the role of collectors.

- Female respondents had firm views on the notion of museums being vital to our culture in the future, with a strong preference for public museums as custodians of science and knowledge, while male respondents were not particularly provoked by any museum-related RDE statements.

- Male respondents believed the geographical focus on Asia would increase, while the female respondents had no strong opinions about future geographical focus.

Age group analysis

Finally, we looked at the different patterns that are identifiable when splitting our respondents into different age groups:

- *18–29 years old*: This is the audience that will make or break the next generation of contemporary artists, the next movements, and particularly the next new media. They gave a mild but nevertheless clear indication that there should be public support for art communities, and showed a negative bias towards collectors, seen as driven primarily by greed.

- *30–39 years old*: Another important sub-cluster, as these are the generation developing the spending power (economic crisis permitting) and the cultural scope to initiate new, important collections, and therefore support new movements, as the 1970s young collectors did with Arte Povera and conceptual art. Almost as a confirmation of this role, this age group demonstrated a strong appreciation of the social aspect of individual artists' success. They also agreed that an artist's individual processes at a personal level will determine the evolution of fine arts, but saw the role of formal academic curricula as key too. This picture of rather orderly organization of the next phase of fine arts was confirmed by the belief that art movements and their manifestos will catalyze the power of individual talents. Even stronger though, was the acknowledgment of the vital public role of art communities and museums: here, the numbers were among the highest of the entire survey. This age cluster positioned museums as *the* custodians of scientific knowledge, and museums in general as a true source of research for business leaders and brand marketers. In the light of the rationale behind this book, this sounds like a very positive data set.

- *40–49 years old*: This is another important group of respondents for market directions and trendsetting, since their disposable income is typically at its highest level. These respondents highlighted the growing role of mass media in recognizing, profiling, and determining which artists are important. In parallel with the 20–29-year-olds, but with a slightly stronger emphasis, this cluster saw the fine arts in future years as being the direct product of the most profound creative processes performed by the artists of today and tomorrow. This is a group that considers the future of fine arts to be strongly biased towards art movements that will act as brands. Perhaps still subscribing to a slightly more conservative vision of the fundamental mechanisms of this sector, they see critics exercising a predominant role in the creation of such movements, in line with the role Germano Celant played for Arte Povera (see Chapter 5).

This group showed the most progressive mindset on the actual content of future fine arts, associating it with urban design and city beautification. This progressive vision of fine arts within society was confirmed by the positive response to statements expressing the value of museums, both at the public level and as a source of inspiration for entrepreneurs and business people. A clear negative bias was provoked by the option of privatizing museums to align them to business management principles. This is an age group that keeps an eye out for the most immediate trends.

- *50+ years old*: These were the most opinionated respondents of the entire sample by far, clearly stating where they thought fine arts was *not* going. On a positive note, they saw art movements as vital frameworks for the future of fine arts, and art communities as vital for the quality of life in cities, both at a generic level and as connecting hubs for the larger communities of the creative class. Additionally, they treasured the experience accumulated by collectors in pursuing their own personal passion as valuable in terms of business learning. For the rest, their position was extremely critical, particularly when addressing the social role and personal dynamics of future artists and their movements. This sub-cluster still believed in the collector as one of the key drivers of future art systems, indeed even with a degree of predominance over museums. In their opinion, the role of public commissioners and public art will not be preponderant in the near future, nor will "design for collectors" be assimilated to the fine arts in distribution and trading. One key detail to close the review of this, the liveliest sub-cluster on both specific questions: these respondents gave a clear indication that they did not believe the "bubble" of the current market would burst or that there would be any future collapse of the fine arts system as we know it today.

These additional "findings" should provide food for thought to fine arts leaders and cultural sector managers. Of course, it cannot be overstressed that the different opinions by gender, age, and nationality will be vital in shaping fine arts in future years. No specific statement should be considered to be *the* essence of the panel, and mining the gold offered by this market research data should be a priority. At the same time, it could be interesting to look at the one statement that enjoyed the highest positive response from everyone. This is revealed next as we wrap up.

Wrapping up: from the market research to a sketch of the future of fine arts

The highest score collectively expressed by our 418 respondents went to the statement regarding "the clear need to support art communities with public funding, in order to keep them alive and kicking during the years of economic difficulty." This confirms the general socially driven orientation of the panel. It is also a sign of hope for a better future, as it highlights a general humanistic awareness among US and UK citizens.

To summarize the overall research experiment:

- We drew out 36 statements, or quantitative research "items," from the interviews with thought leaders and sector experts. They were clustered into six conceptual silos for RDE, on the basis of the last six chapters of this book.

- The respondents were a representative sample (by gender, age, and in general, social status) of US and UK residents (ordinary people, not sector experts). On the basis of their rating of the statements, they were segmented into two groups. The first identified the future of fine arts as "autonomous artifacts," showing a slightly more conservative view, and the second opted for a vision of the future where fine arts will be in the first instance a "social process," displaying a higher degree of alignment with the expert opinions gathered for this book.

- We also analyzed the results by nationality, gender, and age group, to fill out the picture with some more specific findings.

Our respondents did not see the market bubble bursting or the collapse of the entire fine arts system within the next five years as a likely outcome. This is rather encouraging, given the times we are experiencing. Even more encouraging though, is the rather diffused but strongly recurring indication that in one form or another the fine arts and design sectors can offer business leaders and brand marketers a high level of opportunities for learning, exchange, and inspiration. This was most shown most clearly by these responses:

- The status of art communities as cultural R&D labs was validated by the second segment.

- The ability of art communities to connect different fields of the creative industry was validated by both female respondents and the second segment.

- That collectors are potential business leaders thanks to their combination of commercial sense and art sense was validated by the first segment.

- The creative leading role played by museums in helping designers and marketers reflect and improve in their professional domain was validated by UK respondents and by the first segment.

- The essential role played by museums to underpin our culture in general terms was validated by female respondents, by respondents in the key age groups of 30–39 years and 40–49 years, and by the second segment.

- The crucial role that will be played by urban design and city beautification as future directions of evolution for fine arts was validated by the 40–49-year-old age group.

This specific quantitative research project was not designed or carried out to verify the opinions expressed by experts in the qualitative interviews. The purpose was to explore the mindsets and the mood of ordinary people, to look forward with them at what to expect next, while moving towards drafting an algebra of their minds by means of RDE. It was nevertheless reassuring to find that they valued the art and design sectors, saw a strong role for them, and advocated support for them.

What are the next steps? Bearing in mind that our research field was limited to the United Kingdom and the United States owing to budget constraints, the most urgent priority for professionals in the cultural and design sectors should be to extend the breadth and depth of the research by first of all involving more advanced economies that play a vital role in determining tomorrow's fine arts and design (for example, Italy, the Netherlands, the original and expanded European Union), then including emerging countries such as China, India, and Brazil, while not forgetting Japan, South Korea, and other stages where fine arts futures will undoubtedly unfold. A world-wide, world-class project could help to measure the mood and to sketch the directions in which public opinion expects the cultural sector to go, to gain understanding and then make informed decisions – perhaps contrary and radical decisions, but informed ones.

It is appropriate to conclude that (given the indications from a book mainly focused on helping corporate leaders and brand marketers to discover the power of fine arts) the "no sacred cows" principle – as we advocated for commercial business and corporate enterprise – should also apply to the management of culture, design, and fine arts, from here to their next future developments.

A deeper insight into Rule Developing Experimentation (RDE)

RDE is the research methodology resulting from decades of genuine commitment to market research by Howard Moskowitz and described in the bestseller, *Selling Blue Elephants: How to make great products that people want before they even know they want them*, by Moskowitz and Alex Gofman.

Within RDE experiments it is possible to design, test, and flexibly adapt concept ideas and research hypotheses in a scientific way. The particular statistical approach underpinning RDE with proprietary algorithms enables respondents to express their opinions in a practical and efficient way, by means of Internet-based surveys. In particular, it is a feature of the overall methodology to inspire people's expression of visions and solutions even if they do not precisely identify any specific need. This is a classic future studies and predictive market research challenge: every company aims at putting people at the center of their projections. However: how can we express our opinion about future innovations that do not exist yet?

Market research is generally impotent when it comes to predicting future concepts such as the Sony Walkman or the Apple iPod: people's imagination does not encompass products that do not yet exist. RDE addresses this challenge by means of statistical science. Of course, as with all quantitative research, the identification and involvement of the right sample of people to represent the field of exploration is vital: this is where partners such as Peanut Labs are crucial.

During the research online sessions, the panel is presented with a scientifically calibrated number of systematically designed prototypes, or concept ideas. In this chapter, it is possible to see how such concepts were formed, starting, in our case, from the conclusions of thought leader qualitative interviews. The proprietary intellectual property behind RDE then ensures that the full response of people is appropriately treated to provide practical feedback.

RDE has been adopted in projects and in editorial research, by Moskowitz, Gofman, and the author of this book for a prospective publication on the future of premium categories and luxury goods. In addition to its academic soundness and its business effectiveness, it has gained a great degree of popularity all over the world, from the United States to China, because of its simplicity of use and clarity of expression. What better asset to explore the future of fine arts and design in a book otherwise fully committed to a qualitative research approach of an almost journalistic nature?

The world of fine arts, design and culture has a lot to offer in terms of aesthetics, appeal, and even entertainment. It is partly its nature to do so, as beauty and finesse have been intrinsic to its make-up for a long time. Nevertheless, the domains of fine arts and design also have a long track tradition of reflection, inspiration, and even action to achieve change. These have been and still are the territories of crucial battles in societal, ideological, and – of course – political terms. Our journey had the ambition to apply such a natural drive towards change to the thinking and the practices of business in general, and of brand marketing in particular. We started from the self-evident state of deep crisis in our economies and societies. It was not at all the intention to criminalize corporate management or enterprise leaders. However, we built on the assumption that contemporary business models are a clear reflection of the ways of thinking that led us to the present crisis. An ideology of pure capitalism for capitalism's sake as seen in the five years prior to the downturn is the mother of all crises: still now, the ways we work represent an operational demonstration of such a vision. We therefore did our best to discover the context of fine arts and design from various viewpoints, in order to provoke debate and suggest alternative intellectual thinking to what is the frozen fountain of "MBA thinking."

We started our path from fine arts to the art of management with a manifesto. During our journey, we investigated the history, the inner workings, and the present use, of this specific literary format and political tool. I proposed the *Golden Crossroads* manifesto as the basis of this book, in order to present my thesis to the reader. It is time to go back to this manifesto: it is the backbone of the conclusions. These are presented in terms of recapping a few key points for each of the 11 statements originally included in the manifesto. We meet some of the protagonists of this book for last time, and we summarize the 38 findings. To capture the essence of each point, a clear business challenge and business opportunity for entrepreneurs, corporate readers, and brand marketers is suggested in more explicit terms, as an additional comment. This structure offers the valuable feature of being clear and connecting back to my opening statement of intent. I conclude by validating my preliminary ideas in the light of the knowledge we have accumulated so far, going from the first to the last point of the *Golden Crossroads* manifesto.

1 People are the goal, not the means, of any business enterprise that is worthwhile

The reduced financial means and the rediscovered ethic drive of the down-turn years will bring a new people focus. First, it is clear that fine arts and design will recover topics and themes related to social issues. I proposed the example of the Nederlands Fotomuseum in Chapter 6, since this institution is effectively focused on *the* themes of great relevance for the next five years: migration, socialization of minorities, integration of its policies in the area, and more. In parallel, moving to urban trends, we can welcome the practical approach to city beautification by infrastructure design as exemplified by the case of Winy Maas's Flight Forum, in Chapter 2. This general trend was confirmed as part of the opinions gathered at the "Art in the Open Forum 2008" at the ICA in London. This panel, designed with London 2012 in mind, comprised artists, designers, architects, planners, and managers jointly investigating urban futures. Their ideas would not entail the unrealistic pursuit of new dysfunctional icons: the future of urban design lies instead in the aesthetic improvement of vital functional programs.

This is good news from two different, but complementary, points of view. First, the proliferation of rather useless visual icons that affected our cities will likely slow down or even stop, resulting in a reduction of visual clutter. Second, and most important, the new focus of architects, artists, and designers will be on the improvement of what people actually need and use in their lives. How can it not be a better future when beauty is intrinsically designed into those anonymous details that make up our everyday, including our urban infrastructure? What if tomorrow's design restarts from the beautification of those humble functions that make our lives possible? This was the vision of classic Japanese writers such as Tanizaki and Okakura. In the words of the latter: "stop luxury, start the sublime." Will we see in the mid-term future a "sublime" where people stand at the center of the stage? Here, applied arts might manage to retrieve their modern role as a democratic engine of progress.

Challenge

At the time of writing this book, the world of work is still in deep trouble in terms of people management because of the redundancies caused by the crisis. In fall/winter 2008, companies that announced redundancies and layoffs in the thousands were rewarded by a brilliant stock exchange performance the day after. The sheer annihilation of human capital resulted in the financial reward of those entities that caused it. Is it possible to imagine a more perverse situation? This clearly has to change.

Opportunity

It is time to reverse the very fundamentals of corporate capitalism. People should not be viewed as a cost but rather as the only resource that a company actually has to make its future happen. It is simply (and radically) time to go back to humanism as *the* new ideological model to adopt.

2 The field of culture is where reality unfolds: understanding culture(s) is necessary to understand the universe(s) where any business enterprise operates

For too long, business models have been dictated by standardized global manuals, regardless of geographic location or cultural setting. Local cultures deserve much more than just an adaptation of global formats: they deserve instead profound study, at the level of history and in terms of field research. We advocated the necessity to be there, in the field, with integrity and with genuine respect for the local vernaculars. We saw how the business modeling inventions of Davide Quadrio, the founder of BizArt in Shanghai, led to innovative ways of working. We highlighted how rather specialized boutique agencies, such as 515 in Turin, participate in their local cultural milieu, extracting from there assets for their creative leadership. These players are therefore in an ideal position to connect fine arts, design, and corporate communication campaigns. Far better than the heavy corporate "tankers" of large advertising multinationals, it will be for these smaller, highly mobile "wooden ships" of creativity and cultural innovation to redefine what marketing communication will be in the next five years. Getting even deeper into the heart of the system, consulting firms such as Mona Lisa in Paris managed to entrench the power of fine arts creativity, with its "no sacred cows" principle, straight into corporate processes. Building bridges between the cultural universe of artists and the ivory towers of corporate capitalism will be crucial in the near future, in order to help the latter to reform themselves by reinventing the ways they work.

Challenge

In recent months, it has become apparent that just one approach to global management in the post-crisis world will not do any more. Corporate leaders and brand marketers should rethink the role that local cultures and design could play in helping them to change the way they work, at more levels.

Opportunity

Acute agents of culture and commerce such as BizArt, Mona Lisa, and 515 are just a few examples of what innovative boutique companies can achieve by bridging culture and corporate: these firms can plant a different view-point and stimulate the necessary U-turns in the ways we work. This makes change feasible for those who dare to take the first step to pursue it.

3 Culture is everywhere, from the cathedral to the *favelas*: immerse yourself in culture at all levels, by all means

This particular point in the *Golden Crossroads* manifesto was first demon-strated in the choice of research methodologies adopted for this book: in-depth networking within the field, participation, and an "action research" position. Here, marketing research could study the practices of journalistic reporting, in order to gain different views of how knowledge can be gath-ered from the field in a more insightful way. Second, it must be reiterated how studying the history of culture is, and will increasingly be, crucial to understanding countries and market. Reading authors such as Tanizaki and Chatwin might open a thousand golden windows on the cultural mindset of Japanese audiences or on the psychology of collectors in a much more effective fashion than conventional market research actually does.

Challenge

The challenge here is as simple to express as it is spectacular in its mag-nitude. It concerns the need to revolutionize the way business leaders acquire their insights and information about the countries where they operate. This will be feasible if new ways to approach research are used, from field immersions in the actual physical reality of such countries, to a more profound study of their history and mindsets, adopting unconven-tional sources and references.

Opportunity

The above "challenge" is all about the innovation opportunities of market research. The best-case scenario would see market research and corporate leaders moving their understanding of the countries where they operate from that of passive markets to commercially exploit into cultures to creatively learn from, introducing strong elements of co-creation.

4 People's creativity is crucial to business success: be it customer co-design or employee training, nurturing creativity is necessary every time, with any stakeholders

In the future, it will be imperative for corporations, in their capacities as both employers and brand developers, to involve people, moving from yesterday's top-down management, towards tomorrow's co-creative fruitful practices. Creativity has for a long time been the mantra of creative industry giants and corporate brand marketers, but in practice they tend to rely on textbook formulas. In this context, we chose to go deeper in order to understand the mind of the creative better. First, we selected psychoanalysis as our red thread in methodological terms, leaving the floor to Sigmund Freud, C. G. Jung, and Ernst Kris. They studied fine arts and generated a number of hypotheses of great relevance in pure theoretical terms. Inspired by this creative thinking, we demonstrated how these ideas can be made practical thanks to the work of Elliott Jaques on the symbolic dimension of work. We also presented examples of creativity training, from the classic book by James Webb Young to the people coaching practice adopted by Pompei AD in New York. Pompei introduced a yearly training bonus to enable staff members to pursue any activities of their own choice. This put people at the center of the knowledge growth process, by involving each of them on the basis of their personal passions and drive. This seems the way to go in terms of revolutionizing the way we work.

Challenge

The world is divided more than ever into the "haves" and "have nots." With systematic staff reductions by corporations and companies in general, the workplace has been a place of uncertainty and fear for quite some time now. Reversing this state of affairs should be *the* priority for business leaders, everywhere: it is to their advantage in terms of both the quality of output of their teams, and a rediscovered confidence in their markets.

Opportunity

Profound engagement starts from true debate: here, management have a golden chance to look at their staff not as just numbers but as powerful agents of change. The mindset of corporate leaders has mostly been focused on rationalization measures and financial indicators. Beyond sterile engagement surveys that have no follow-up, it is time to install a democracy index in companies, to measure the degree of co-decision making, and make it a crucial performance indicator of top management.

5 Stop producing neutral PowerPoint presentations, start generating ideologies, and a vision of the world as it should be, according to you

I reiterate here once again my belief that the nature of the present crisis is ideological. The creation of new ideologies is a necessity to redefine our starting point towards the next steps to emerge from today's difficulties. The format of manifestos was identified as a practical format to launch new visions of the world and to impact wider audiences thanks to practical synthesis, verbal eloquence, and conceptual clashes. We studied how Marinetti used the manifesto format to create Futurism from scratch. We discovered how Platform 21 is redefining the ideological profile of Dutch design by means of its Repair Manifesto. The fact that the essence of this very book is captured in a manifesto is the best testimonial to my strong confidence in this particular form of communication.

Challenge

Once upon a time, brand design was all about vision, big ideas, and the power of rhetoric to launch new propositions based on a unique asset. In time, corporate branding became increasingly a place of sterile formulas for prompt production. What if the power of fine arts' modern history and thinking could help revert this?

Opportunity

Here, I have just one only hope: that one, ten, hundreds of manifestos will be generated by corporations and enterprises to redesign not only their business standpoints, but also their new role in society as active forces of positive change.

6 Stop producing products, start generating stories, and a narrative line that boldly tells the rest of the world who you are

From the great manifestos of modern art and contemporary design, we turned our focus to the realm of tangible artifacts. We went back to the 1950s classic semiotic analysis of the Citroen DS by French thinker Roland Barthes to achieve a clear vision of how products are sociological artifacts that generate sense in our collective stories. Because storytelling and viral debates are the places where culture happens, this is *the* domain to be addressed by tomorrow's brand marketing and product innovation to achieve long-term success. Here, the art marketing track record of Seth Siegelaub is an important reference. By packaging conceptual art into appealing market offerings, his experience shows the fundamental role that propaganda plays when rightly attuned to tell the story. Because ultimately, a brand is an idea in people's minds, and a product, especially within lifestyle categories, is only as good as the narrative line that underpins its very existence.

Challenge

In the late 1990s, advertising went through a major revolution thanks to the switch from mass media to digital media. This meant both a generational change at the very heart of the industry in terms of creative direction and a profound process of rethinking the very fundamentals of the business. The same challenge now awaits chief marketing officers and brand strategists across categories and industries: the current state of affairs calls for a redefinition not only of what brands ideologically are, but also of the ways they reach people.

Opportunity

The opportunity exists to go through this rebirth process of soul searching and coherent narration of what your enterprise and its brands stand for. Once again, it is boutique agencies and atypical players in the consulting arena who could be best positioned to inject the creative sparkle into corporate processes, and facilitate change.

7 Stop producing products, start enabling and supporting genuine communities, with the greatest integrity and spirit of service

One of our quantitative segments identified future fine arts as a socially attuned process. This also happened to be the most active segment contributing to the RDE work. Additionally, according to the majority of respondents across the entire survey, the existence of communities in the fine arts and in the design worlds should be protected, even to the point of supporting them by public funding, if necessary. We might guess from this statistic that such communities are where people's interest will be in the future. What a great opportunity for companies to take the lead and start profiling themselves as enablers of communal life in cities both large and small! Examples of how fine arts players benefited from a genuine commitment to community building abound. We presented Nakaochiai Gallery in Tokyo, managing to socially bond with the population of a non-arty, non-trendy downtown area through art; and we introduced Tokyo Gallery, managing to expand its cultural and commercial reach to Beijing, against all the odds, after connecting South Korean artists to the Japanese scene in the 1970s. Art communities can be seen as tomorrow's true experimental lab where new movements are born, new ideas are circulated, and the natural thirst for the new by collectors is met by new offerings. In essence, between marketplace and social hub, communities will be among the drivers of tomorrow's fine arts systems, and beyond.

Challenge

Business management generally tends to think in terms of short-term performance indicators, with people seen purely as marketing targets. This is a major cultural roadblock to truly understanding the power of community endorsement within marketing teams. What about reaching people's hearts instead of just tracking them in supermarkets?

Opportunity

The great demand for community building in cities is a long-term trend: fear and uncertainty will bring people to gather and to connect. These are not necessarily customers who want to buy your products now. They however will potentially be tomorrow's customers. Their communities are where you find their interests and their drives. Especially in times of down-turn, turning strategic attention and marketing dollars to this societal field will mean conquering people's hearts. There is only one condition: here, your integrity and genuine commitment are a must.

8 Design your brand for the long term: there are no tactics, take every decision as if its impact was eternal

We see design as a crucial constituent of our culture, to the same degree as fine arts and architecture, but with different scope and objectives. The power of design is to capture the essence of a brand and make it immediately visual. Such capability sometimes translates into opportunities for strategic consulting, as performed by Thonik for the Centraal Museum of Utrecht. It also leaves space to challenge traditional corporate identity philosophies, as the Van Abbemuseum art direction across the decades did. Both a mature contributor to business success and a maverick *enfant terrible* of corporate organizations, design requires the integrity and the ethical commitment always to investigate reality before making it seductive. The ideological and intellectual limits of contemporary design, as highlighted by Hal Foster in his 2002 essay *Design and Crime*, too often made it into an acritical partner in crime to business. It is time to change this, and to find a new critical – or even radical – edge to design, in order to ensure that its relevance goes beyond what Munari defined as "styling," and back into creative thinking, where it belongs.

Challenge

Design as a gimmick or as an expensive add-on will not be *the* direction of development for this discipline in the next five years or so. Recent years saw a rapid proliferation of design for the benefit of both industry and commerce. This led to a notion of design as gimmick that will definitely not grow further. Companies should take notice and change design strategies accordingly.

Opportunity

The magical ability to connect design with society and business will be vital in the future: this is where tomorrow's socially sensible icons will be generated. Corporate design teams are a concentration of world-class talent: it is time to utilize even a small amount of their capacity to redesign what design is. This strategic investment would pay off with midterm success on the cultural circuit, with excellent potential media and reputation benefits.

9 Design your brand as an incomplete universe: holistic in ambition, co-creative in people's participation, utopian in its reach

It was apparent from our analysis that in the longer term people will think of the pervasive presence of rigidly controlled images of lifestyle and luxury brands as dysfunctional. In reaching this conclusion, we established a theoretical line of foundation for our argument that goes from Loos to Foster, from the early 1900s to the early 2000s. The necessity to rethink what your brand is in society and where your enterprise is going in political terms calls for a new and holistic vision at the abstract level of creative thinking. In parallel, the demand from people every day is to open corporate design and creative processes to genuine participation and creative debate. This is a complex game of opposing stimuli, where the best talent will be needed to define and use new forms of design leadership.

Challenge

Combining the encompassing vision of a new cultural profile with the necessity to leave the floor to design users will be the challenge of each company generally but in particular of lifestyle and luxury blue-chip companies. Here, the perfect storm conditions see the merging of the financial and economic impact of the crisis with a historical drive to top-down art direction, contrary to current cultural trends.

Opportunity

It is for luxury and lifestyle brands to reverse this dangerous situation and take the lead. They can learn a lot from the classic and modern history of fine arts in terms of what style and signature as brand can mean to them. Talent is paramount in these companies: it is time to rearrange it in a different way to create new value.

10 Design and maintain the integrity of your brand according to your own vernacular grammars, and let your corporate identity flexibly follow your evolution over time.

In the near future, consistency and coherence will not just mean the dictatorship of identity design manuals and corporate guidelines. They will instead mean a great degree of "soul searching" and the ability to go back to your roots, to regenerate them. We saw how the contemporary history of Dutch design recorded a move from somewhat antagonistic aesthetics to vain commercialism, at the risk of losing cultural relevance and commercial sustainability. It will take time collectively to fix such a status quo. Companies should recapture their own "cultural DNA" in terms of ideology and ensure that all steps are planned over time, to build an authentic ethos. Not doing so will expose brands and corporations alike to the deadly combination of weak demand and sharp socio-cultural scrutiny by consumers worldwide.

Challenge

Soul searching is not an easy exercise. It means being ready to confront your own organization with its history, its roots, and especially its deviations from its original position. This is the kind of exercise that sometimes leads to long and painful sessions of self-analysis and rediscovery.

Opportunity

It is my direct experience that using brand design from an organizational viewpoint is an excellent opportunity to conduct this kind of "soul check," and to reinvent your company both as a brand, and as a living organization. There is no doubt that the marketplace will not grant any sympathy to those brands that are not true to themselves: what might seem like pain is just rebirth.

11 The world of fine arts, design, and culture is important because it is different and diverse: always nurture difference and diversity in every way you can

Artists, primary galleries, auctions, dealers, and collectors: throughout this book, we discovered a world of characters, where everybody is a protagonist in their own way, particularly in contributing to the rich texture of the fine arts and design worlds. The fact that a consistent segment from our quantitative study still sees the future of fine arts as very much driven by the making and selling of autonomous artifacts is the best testimonial to the cultural relevance of this sector, even in these times of economic crisis. We might say that from the pathological genius as exemplified by the sculptor F. X. Messerschmidt to the entrepreneurial mindset of contemporary masters such as Damien Hirst and Takashi Murakami, artists themselves comprise a huge reservoir of diversity at all levels. Even more encouraging, the huge American and Japanese success of Emily Kame Kngwarreye, the late Australian aboriginal amateur painter become a world star, indicates that the art system is still open to including and endorsing different approaches, original visions, and the power of talent – even if it does not emerge from an orthodox academic curriculum.

At the other end of the spectrum, we discussed the case of the Chinese creative industry and its applied arts. Here, the absence of real contrarians and the systematic rejection of critical thinking at a national didactic level affect the ability to perform at international creative industry standards. Our natural conclusion is that the creative practices and the applied arts will always need those who do not agree, those who do not join the chorus, and those who do not want to fit. It is this openness that makes the art, design, and cultural sectors vibrate with talent and resonate with societies: finally, in spite of the business and academic sides to them, these are still domains where the unconventional, atypical practitioners, and the irregulars of this world can find not only a place, but their actual path to importance and success, if they have a real talent.

Challenge

"Diversity" and "inclusion" are typical keywords of corporate mantras today. The way they are executed though, especially in the light of the downturn, points much more towards the mechanical aggregation of people of diverse origin around the same very limited set of sterile values and norms. This is pursued as a top-down HRM strategy instead of creating an open space for creative differences from such sterilizing standards. Although expressed as initiatives with some degree of good intentions, these programs tend inevitably to slip into the lobbying of particular minorities to gain more corporate power positions, and not much more. This will certainly not be the way forward.

Opportunity

In the 1970s and 1980s, it was a badge of honor of the advertising industry to attract the most diverse personalities and to nurture them by creating professional environments sometimes bordering on the extravagant. Although this sometimes led to excess, this drive to accept real diversity in terms of behavior, of values, and of ideas should be regained as soon as possible by a corporate sector that has destroyed true differences within its ranks. The opportunity here is simple and huge at the same time: stop looking at your culture, at your organization, and at your brand as fixed entities. Start looking at them instead as living organisms, in need of both emotional and ideological complexity. Then redesign your HRM policies accordingly. That will be the first step towards creative edge and a new supremacy of true talent.

Now that I have recapped all 11 points of the manifesto in the light of content from the nine chapters of this book, the natural conclusion is to offer a practical suggestion to business readers. What really matters today is not academic depth or rigid systems. What matters most is inspiration and action: making things happen in the field, changing the world, and ultimately experimenting with your professional and personal life. These are the vital traits of the creative industry and fine arts at their best. This is why I do not offer *the* way to open out and to implement the ideas in this book in the sense of just another toolkit.

This book was not conceived as a "how to" manual, and accordingly it does not vulgarize the complexity of the challenges ahead of us all, and of the opportunities available to those who will dare to change. Instead of doing this, I suggest that business leaders and brand marketers, and any other reader with goodwill and a real passion for change, should select a few ideas, some specific cases and stories, or even just the one single point of the manifesto, and then act on it as soon as possible and as profoundly as is feasible.

Fine arts and design are the realms of what is produced, not of what is theoretically described. Likewise, as much as I believe in each of the words printed in this book as being relevant in the current context, what ultimately makes the difference is not how well you memorize them or how much you enjoyed reading them: it is what you will do with these humanistic visions and with the line of optimistic hope that binds this book together. Once again, please make your own selection of what you feel truly mattered to you in this book. Then, make sure to take immediate action by applying it to the real world you live in, to make it a better world. This was my aim, this is my last call for action, with the hope that some readers will reach a position to provide me with updated "findings" in the next editions of this book.

Alberro, A. (2003) *Conceptual Art and the Politics of Publicity*, Cambridge, Mass./London.

Alpers, S. (1988) *Rembrandt's Enterprise. The Studio and the Market*, Chicago.

Apollonio, U. (1970/73) *Futurist Manifestos*, London.

Baghar, M., Cooley, S. and White, D. (2000) *The Alchemy of Growth,* New York.

Barthes, R. (1957) *Mythologies,* Paris.

Betsky, A. (2004) *False Flat: Why Dutch Design is So Good,* London.

Bevolo, M. (1994) *The Art of Advertising*, Milan, Italy.

Bevolo, M. (2007) *City, People, Light*, Eindhoven, Netherlands.

Bevolo, M. et al. (2010) *Premium by Design*, Wharton School Publishing, USA.

Boekraad, H. (2008) *Power: Thonik Design*, Shanghai, PRC.

Bradbury, H. and Reason, P. (eds) (2001) *Handbook of Action Research*, London.

Branzi, A. (2006) *Modernità debole e diffusa*, Milan, Italy.

Calvi, G. (1972) *Nuove questioni di psicologia: La Creatività*, Brescia, Italy.

Chasseguet-Smirgel, J. (1971) *Pour une psychoanlyse de l'art et de la creativité*, Paris.

Chatwin, B. (1989) *Utz*, London.

Cole, A. (ed.) (2007) *Design and Art*, London.

Debbaut, J. and Verhulst, M. (2002) *Van Abbemuseum: Het Collectieboek*, Eindhoven, Netherlands.

Depero, F. (1930/40/90) *Un Futurista a New York*, Montepulciano, Italy.

Eichhorn, M. (2009) *The Artist's Contract*, Cologne, Germany.

Ferrari, F. and Ferrari, N. (2006) *The Furniture of Carlo Mollino*, London.

Foster, H. (2003) *Design and Crime*, London/New York

Freud, S. (1920) *A General Introduction to Psychoanalysis*, New York.

Freud, S. (1969/71) *Saggi sull'arte, la letteratura ed il linguaggio*, Turin.

Gofman, A. and Moskowitz, H. (2007) *Selling Blue Elephants*, Upper Saddle River, N.J., USA.

Gordon Cantor, S. (2002) *Alfred H. Barr, Jr. and the Intellectual Origins of the Museum of Modern Art*, Cambridge, Mass./London.

Hata, K. (2004) *Louis Vuitton Japan: The Building of Luxury*, New York.

Hauser, A. (1955) *Sozialgeschichte der Kunst und Literatur*, Muenchen, Germany.

Hirst, D. (2007) *The Making of the Diamond Skull*, London.

Hitchcock, H. R. and Johnson, P. (1932) *The International Style* (Catalogue of the MoMA exhibition), New York.

Jaques, E. (1970) *Work, Creativity and Social Justice,* London.

Jung, C. G. (1907) *Der Dichter und das Phantasieren*, Zuerich.

Koolhaas, R. and Mau, B. (1995) *S, M, L, XL,* New York.

Kris, E. (1952) *Psychoanalytic Explorations in Art*, New York.

Loos, A. (1908) *Ornament and Crime*, tr. English 1997 by Ariadne Press, California.

Lugli, A. (1992) *Museologia*, Milan.

Marinetti, F. T. (1916/2003) *Come si seducono le donne*, Florence.

McAndrew, C. (2009) *Globalization and the Art Market*, Helvoirt, Netherlands.

Melucci, A. (1994) *Creatività: Miti, Processi, Discorsi*, Milan.

Mienke, S.T. (2008) *Dutch Design: A History*, London.

Munari, B. (1971) *Artista e Designer*, Bari, Italy.

Murakami, T. (2000) *Post Pop = Superflat*, Tokyo.

Nicholson, G. (2006) *Sex Collectors*, New York.

Okura, K. (1913) *The Book of Tea*, Tokyo.

Perloff, M. (1986) *The Futurist Moment*, Chicago.

Pomian, K. (2003) *Des saintes reliques à l'art moderne*, Paris.

Previeux, J. (2008) *Lettres de non-motivation*, Paris.

Rawlings, A. (ed.) (2008) *Art Space Tokyo*, Tokyo.

Reeves, R. (1961) *Reality in Advertising,* New York.

Rothko, M. (2004) *The Artist's Reality*, New Haven, Conn./London.

Santagata, W. (1998) *Simbolo e Merce*, Bologna, Italy.

Schubert, K., (2000) *The Curator's Egg*, London.

Schumpeter, J., (1942) *Capitalism, Socialism and Democracy*, new edn, 2008, London

Slaughter, R. A. (2004) *Futures beyond Distopia*, London.

Smith, K., (2005) *Nine Lives: The Birth of the Avant-Garde in New China,* Zurich.

Sudjc, D. (2008) *The Language of Things*, London.

Tanizaki, J. (1935) *In'ei Raisa (In praise of shadows),* Tokyo, tr. Italian 2002, Milan.

Thompson, D. (2008) *The $12 Million Stuffed Shark*, London.

Vars authors (1991) *Flash Art: XXI Anni*, Milan.

Vars authors (1991) *Kunst + Design: Wim Crouwel*, Stuttgart, Germany.

Vars authors (2001) *Thonik*, Rotterdam, Netherlands.

Vars authors (2005) *Mary Vieira: O Tempo do Movimento*, Sao Paulo, Brazil.

Vars authors (2005) *Yoshio Taniguchi: Nine Museums*, New York.

Vars authors (2008) *Tokyo Chanel Mobile Art Issue # 2* (catalogue), Tokyo.

Vars authors (2008) *Culture and the Human Body: Prince Claus Awards 2008* (catalogue), Amsterdam, Netherlands.

Vars authors (2009) *TEFAF Maastricht 09* (catalogue), Helvoirt, Netherlands

Vettese, A. (1991) *Investire in Arte*, Milan.

Young, J. W. (1940) *A Technique for Producing Ideas*, Chicago.

Articles and chapters

Celant, G. (1967) "Arte Povera: appunty per una guerriglia," *Flash Art* 5, Nov./Dec., Rome.

Deitch, J. (1980) "Arte come investimento," *Flash Art* 94/95, Jan./Feb., Milan.

Depero, F. and Balla, G., (1915) "Futurist reconstruction of the universe," reprinted in *Apollonio*, 1970.

Online resources

Cannell, M. (2009) "Design loves a depression," *International Herald Tribune*, January 25, available at: http://www.nytimes.com/2009/01/04/weekin review/04cannell.html?_r=4&scp=1&sq=lasky&st=cse